Alternative Approaches to Causation

Alternative Approaches to Causation

Beyond Difference-making and Mechanism

Edited by

YAFENG SHAN

OXFORD
UNIVERSITY PRESS

Great Clarendon Street, Oxford, OX2 6DP,
United Kingdom

Oxford University Press is a department of the University of Oxford.
It furthers the University's objective of excellence in research, scholarship,
and education by publishing worldwide. Oxford is a registered trade mark of
Oxford University Press in the UK and in certain other countries

Published in the United States of America by Oxford University Press
198 Madison Avenue, New York, NY 10016, United States of America

British Library Cataloguing in Publication Data
Data available

Library of Congress Control Number: 2023944515

ISBN 978-0-19-286348-5

DOI: 10.1093/oso/9780192863485.001.0001

Printed and bound by
CPI Group (UK) Ltd, Croydon, CR0 4YY

Links to third party websites are provided by Oxford in good faith and
for information only. Oxford disclaims any responsibility for the materials
contained in any third party website referenced in this work.

Contents

Acknowledgements

The editor would like to thank all the contributors and referees for their hard work on this volume and Peter Momtchiloff and Tara Werger at Oxford University Press for their work in bringing the book to fruition. The editor is grateful to Kok Tsun (Anson) Liu for his hard work on indexing. The editor would also like to thank the British Academy for its support (SRG1920\101 076).

List of Contributors

Rani Lill Anjum, Faculty of Environmental Sciences and Natural Resource Management (MINA) and the School of Economics and Business, Norwegian University of Life Sciences, Norway

Antony Eagle, Department of Philosophy, University of Adelaide, Australia

Evan Fales, Department of Philosophy, University of Iowa, USA

Phyllis Illari, Department of Science and Technology Studies, University College London, UK

R. D. Ingthorsson, Department of Philosophy, History and Art Studies, University of Helsinki, Finland

John D. Norton, Department of History and Philosophy of Science, University of Pittsburgh, USA

Julian Reiss, Institute for Philosophy and Scientific Method, Johannes Kepler University Linz, Austria

Elena Rocca, Department of Life Sciences and Health, Oslo Metropolitan University, Norway

Federica Russo, Freudenthal Institute, Utrecht University, The Netherlands

Yafeng Shan, Division of Humanities, Hong Kong University of Science and Technology, Hong Kong

Samuel D. Taylor, Department of Philosophy, University of Kent, UK

Michael Tooley, Department of Philosophy, University of Colorado, Boulder, USA

Jon Williamson, Department of Philosophy, University of Kent, UK

1

Some Reflections on Causation

Yafeng Shan

1. Recent Debates over Causation

Philosophical analyses of causation have been centred on the question of what causation is. From a philosophical point of view, this question can be interpreted in at least four different senses.

Metaphysical issue: What is causation *out there*?

Epistemological issue: How can a causal claim be established and assessed?

Conceptual issue: What does the word 'cause' mean?

Methodological issue: What methods ought one to use in order to establish and assess a causal claim?

A popular way to explore the nature of causation is to analyse the concept of causation in terms of more fundamental, non-causal notions. Much effort has been made to explicate what these fundamental, non-causal notions are. It has been widely debated whether causation is reducible to regularity (e.g. Mackie 1974; Graßhoff and May 2001), probabilistic dependence (e.g. Reichenbach 1956; Suppes 1970; Glynn 2011), counterfactual dependence (e.g. Lewis 1973; 2000), disposition or power (e.g. Harré and Madden 1975; Anjum and Mumford 2010; Bird 2010), intervention (Woodward 2003), mechanism (e.g. Glennan 1996; Machamer, Darden, and Craver 2000; Machamer 2004), transmissional process (e.g. Salmon 1980; Kistler 1998; Dowe 2000), information transfer (Collier 1999), or something else. Most of these accounts of causation can be classified into two approaches: the difference-making approach and the mechanistic approach. The regularity account, the probabilistic account, the counterfactual account, and the interventionist account can all be characterized as variants of the difference-making approach to causation, whose basic idea is that a cause is what makes a difference. In contrast, the mechanism-based account, the transmissional process-based account, and the information transfer-based account can be construed as variants of the mechanistic approach, which assumes that two events are causally connected if and only if they are connected by an underlying mechanism of some

Yafeng Shan, *Some Reflections on Causation* In: *Alternative Approaches to Causation: Beyond Difference-making and Mechanism.* Edited by: Yafeng Shan, Oxford University Press.

appropriate sort. In the past two decades, the difference-making and mechanistic approaches somehow dominated the philosophical examination of causation, especially in the philosophy of science. It has been assessed whether they can be fruitfully applied to causal enquiry in the natural and social sciences (e.g. Russo 2009; Chao, Chen, and Millstein 2013; Reutlinger 2013; Runhardt 2015; Zwier 2017; Maziarz 2020).

That being said, there have still been attempts to explore alternative approaches to causation. Causal pluralists (e.g. Anscombe 1971; Cartwright 2002; Hall 2004; Crasnow 2011; Illari and Russo 2014; Reiss 2015) doubt that any one variant of the difference-making or mechanistic approaches can fully capture the nature of causation, given the diversity of the use of causal concepts in different contexts. It has been argued that there is a plurality of types or concepts of causal relationships. Evidential Pluralism was also developed to accommodate the fact that there are a variety of methods to establish and assess causal claims across the sciences. In contrast to causal pluralists, Evidential Pluralists do not think that there are different types or concepts of causal relationships. Rather they argue that there are just different types of evidence obtained from various methods for a causal claim. The key idea of Evidential Pluralism is that in order to establish a causal claim one normally needs to establish the existence of an appropriate correlation and the existence of an appropriate mechanism complex. It has also been argued that Evidential Pluralism can be applied to the biomedical and social sciences (e.g. Russo and Williamson 2007; Gillies 2011; Clarke et al. 2014; Wilde and Parkkinen 2019; Maziarz 2021; Shan and Williamson 2021; 2023). Causal non-reductionists (e.g. Tooley 1987; 1990; Carroll 1990; Fales 1990) argue that causation is primitive and unanalysable, while causal eliminitivists (e.g. Russell 1912) maintain that science has no need of the concept of causation at all. Recently, some traditional approaches, especially the dispositionalist approach (e.g. Mumford and Anjum 2011; Anjum and Mumford 2018), were further developed in a more science-informed way, whereas new approaches, including the epistemic approach (e.g. Williamson 2005; 2006a; 2006b), the fictionalist approach (Eagle 2007), the inferentialist approach (Reiss 2012), and the powerful particulars approach (e.g. Ingthorsson 2002; 2021), were introduced. Unfortunately, these approaches, though important and promising, are to a great extent inadequately discussed and assessed.

This book explores and examines these alternative approaches to causation. It revisits causal non-reductionism and causal eliminativism in the context of recent literature. It further explores the pluralistic approach, the fictionalist approach, and the inferentialist approach. It also examines the application of the dispositional approach, the epistemic approach, and the powerful particulars approach to the natural and social sciences.

2. Alternative Approaches: Beyond Difference-making and Mechanism

An obvious alternative to the difference-making and mechanistic approaches is the so-called non-reductionist approach, sometimes also called primitivism or anti-reductionism. The central thesis of the non-reductionist approach is that causation is non-reductionist or primitive in two senses:

a. Causation is not supervenient on the total history of the world and laws of nature.
b. Causal relations are not reducible to other states of affairs, including non-causal properties or relations.

Therefore, from a non-reductionist's point of view, the concept of causation cannot be analysed in terms of difference-making or mechanism. As a leading proponent of the non-reductionist approach, Michael Tooley provides a detailed and systematic defence of a non-reductionist, theoretical-term account of causation in Chapter 2. He argues that such an account of causation has some highly desirable properties: it is by far the simplest of all analyses of causation; it immediately entails the correct relationships between causation and probabilities; and it provides the basis for the only satisfactory account of the relation of temporal priority. Then he examines three alternative accounts: Humean reductionist approaches (especially variants of the difference-making approach), non-Humean reductionist approaches (such as the dispositionalist approach), and the view that the concept of the relation of causation is analytically basic (see a defence of this view in Chapter 3). Tooley argues that all of those approaches are open to some decisive objections, so none of them pose any serious challenges to his non-reductionist, theoretical-term account of causation.

Evan Fales, another long-term supporter of the non-reductionist approach, defends a different version of non-reductionism—namely, robust causal realism—in Chapter 3. Although he agrees with Tooley that causal relations are not reducible to other states of affairs, Fales contends that the concept of causation is analytically basic and thus unanalysable. His central argument is that causal relations are basic and in some instances directly perceived; they cannot be analysed by way of subjunctives, dispositions, regularities, or the like. He takes causal relations to be grounded in a second-order non-contingent relation between the fundamental properties that ground (many) laws of nature. In the chapter, he examines two (related) issues, one concerning the epistemological implications of the view (specifically, its support for inductive inferences), and the other ontological: viz. exploring the nature of the basis for the relationship between a universal's nature and the causal relations it bears to other universals.

Even if they are interested in the epistemological issues, both of Tooley's and Fales's concerns are fundamentally about the metaphysics of causation (i.e. what is causation out there?). However, John Norton is highly sceptical of such a kind of enquiry. In Chapter 4, he makes an empirically based, sceptical critique of the metaphysics of causation. Norton presents an old dilemma for any metaphysical analysis of causation:

> EITHER conforming a science to cause and effect places a restriction on the factual content of a science; OR it does not.[1]

If the metaphysical analysis of causation aims at some pronouncements on the nature of causation as a deep metaphysical truth, history shows us a sustained record of failure: many pronouncements are shown to be false with the development of science. If the metaphysical analysis of causation adds nothing factual to what we already know by empirical science, it is difficult to see what its value is. Nevertheless, it should be noted that Norton does not try to dismiss the value of any philosophical analysis of causation at all. He argues even if designations of causality are mere matters of definition, some designations are worthwhile. And the interventionist account is such a designation. In a nutshell, Norton argues that the identification of causal processes in the world is one not of factual discovery, but of the application of convenient definitions.

Norton's conclusion might seem polemic, but his message is clear and loud: any serious philosophical examination of causation ought to take science seriously. Chapter 5 by Rani Lill Anjum and Elena Rocca is an exemplar of engaging the philosophical analysis of causation closely with scientific research. They develop a traditional metaphysical approach—namely, the dispositionalist approach (or Causal Dispositionalism, to use their term)—and try to show how it can inform and improve the practice of the biological sciences. According to Causal Dispositionalism, causes are intrinsic dispositions or powers, which cannot be further analysed into other states of affairs.[2] Importantly, these dispositions or powers can be unmanifested. Accordingly, a causal process typically results from the interaction of multiple dispositions in some appropriate circumstances. They contrast Causal Dispositionalism with the Humean two-event model, according to which causation is an unobservable relation between two observable occurrences. They argue that the well-known problem of low external validity in medical research, gene technology, and ecological study is a product of standard risk assessment approaches used in these fields that are motivated by the Humean

---

[1] This dilemma was first posed by Norton in an earlier paper (Norton 2003) and is refined in Chapter 4.
[2] This is why for Causal Dispositionalists (e.g. Stephen Mumford, Anjum, and Rocca) causation is primitive, though in a different sense compared with the non-reductionist account of causation.

two-event model. Anjum and Rocca call for a change of the methods for studying causal complexity and interactions in standard risk assessment by replacing the Humean two-event model with the dispositionalist model as the underlying, implicit philosophical assumption.

R. D. Ingthorsson argues for his powerful particulars approach in Chapter 6. In contrast to the difference-making and mechanistic approaches, the powerful particulars approach maintains that causal production is a reciprocal interaction rather than a unidirectional process. Ingthorsson critically assesses whether four rival accounts of efficient causation—the transmissional process-based account, mechanism-based, dispositionalist, and powerful particulars accounts—can well characterize the scientific understanding of two often discussed cases of causal phenomena: (i) collisions between billiard balls and (ii) how water dissolves salt. He argues that only the powerful particulars view can be considered compatible with our scientific understanding, mainly because the other three characterize interactions—to varying degrees—in terms of a unidirectional exertion of influence of one thing on another, which is incompatible with the established scientific understanding that all interactions are perfectly reciprocal.

Phyllis Illari and Federica Russo put forward a different account (viz. the information transmission account) of causal production in Chapter 7. It is worth highlighting that this information transmission account should be understood within their pluralist framework of causation in general, what they call 'the causal mosaic' (Illari and Russo 2014). Illari and Russo are causal pluralists in the sense that they regard different approaches, such as in terms of counterfactuals, dispositions, and mechanisms, as tiles that need to be put next to one another to form a dynamic image of the mosaic of causal theory. They highlight the role of causal production within the causal mosaic, and argue that a metaphysics for causal production as information transmission is in need of a set of epistemological strategies to trace whether information is transmitted or not; such epistemological strategies can be in terms of 'variation' as well as 'information transmission'. They also explain the 'ontoepistemological' combination of information transmission and variation with the aid of four examples of causal production.

In contrast to other authors, Julian Reiss focusses on the conceptual issue of causation. He poses a challenge to a standard way of analysing the concept of causation in philosophy, which is neatly summarized by John Carroll:

> Philosophers routinely seek a certain sort of analysis of causation. They have sought a completion of
>
> (S) *c* caused *e* if and only if . . . (Carroll 2009, 279)

Reiss argues that none of the philosophical accounts which aim to provide truth conditions for causal claims like (S) can be a fully general theory of causation. In

Chapter 8, Reiss argues for an inferentialist approach, according to which the meaning of a causative statement is given by the network of statements with which it is inferentially related. He contends that the inferentialist account of causation is the only theory of causation that can handle the richness of our causal vocabulary satisfactorily.

In Chapter 9, Antony Eagle defends a fictionalist approach to causation. He argues that the models of fundamental physics do not involve any obvious role for causal relations. However, Eagle admits that causation is nevertheless pragmatically indispensable, especially in the human sciences. He thus argues that while we ought to disavow belief in causes we ought to retain causal talks and our deployment of causal models in situations of human interest.

In Chapter 10, Yafeng Shan, Samuel Taylor, and Jon Williamson develop the epistemic approach to causation, which views causation as a tool that helps us to predict, explain, and control our world, rather than as a relation that exists independently of our epistemic practices. They outline four considerations that motivate the epistemic theory. First, they argue that the epistemic approach does not succumb to the problems and counterexamples that beset difference-making and mechanistic approaches and causal pluralism (the argument from failure). Second, they argue that the epistemic approach is simpler than its rivals as it does not need to analyse causality in terms of one of its indicators, or in terms of a pluralist panoply of indicators (the argument from parsimony). Third, they argue that the epistemic approach can accommodate Evidential Pluralism perfectly well, while Evidential Pluralism poses a serious challenge to standard accounts of causation (the argument from Evidential Pluralism). Fourth, they argue that the epistemic approach is needed to provide a unified conception of causal explanation in cognitive science (the argument from neutrality). They illustrate these four considerations in the contexts of the social sciences and the cognitive sciences. They conclude that the epistemic approach provides a very natural account of causation across these contexts.

3. A Practice Turn (and Towards a Functional Approach?)

As we may notice from some chapters of this book as well as the recent literature, there is a gradual, ongoing practice turn in the philosophical examination of causation. More and more philosophers are paying serious attention to methods of causal enquiry across the sciences and exploring the philosophical implications. There has been an increasing interest in the methodological issue of causation. For example, the role of Randomized Controlled Trials (RCTs) in causal enquiry in the biomedical sciences has been extensively evaluated (e.g. Cartwright 2007; 2010; Worrall 2010; Northcott 2012; Krauss 2021), while the legitimacy of process tracing in the assessment of causal claims in political science has been hotly debated (e.g. Crasnow 2017; Beach 2021; Runhardt 2021; Dowding 2023).

What is more, philosophers have begun recognizing the significance of the use of causal claims in practice. In addition to the metaphysical, epistemological, conceptual, and methodological issues, the practical issue of causation is now being taken seriously.

Practical issue: What is a causal claim for in practice? Or, what is the use of a causal claim?

In this book, Illari and Russo regard the practical issue as one of the five important philosophical questions of causation. Eagle highlights the instrumental feature of causal talk in the human sciences. Shan, Taylor, and Williamson emphasize the role of causal claims in predicting, explaining, and controlling the phenomena in the world. All these seem to echo the so-called functional approach, proposed by James Woodward.

[B]y a functional approach to causation, I have in mind an approach that takes as its point of departure the idea that causal information and reasoning are some-times useful or functional in the sense of serving various goals and purposes that we have. It then proceeds by trying to understand and evaluate various forms of causal cognition in terms of how well they conduce to the achievement of these purposes. Causal cognition is thus seen as a kind of epistemic technology—as a tool—and, like other technologies, judged in terms of how well it serves our goals and purposes. (Woodward 2014, 693–94)

I particularly like Woodward's idea of taking 'causation' as a tool. As I have argued, a scientific concept should be understood as a tool for scientists to define and solve problems in practice (Shan 2020, 147). More specifically, I argued that a scientific concept is used as a tool to contribute to various intertwined activities in scientific practice, including problem-defining, hypothesization, experimentation, and reasoning. We should understand conceptual practice in a broader context of scientific practice. Accordingly, regarding causal concepts, we need to analyse and examine them within the context of scientific practice by scrutinizing how a causal claim contributes to various scientific activities.

Moreover, I argue that the practical issue of causation is highly relevant to the metaphysical, epistemological, conceptual, and methodological issues of caus-ation. In contemporary discussion of causation, the ultimate task for philosophers is to develop an approach to articulate and account for the nature of causation. There is a philosophical motivation for such an enquiry: we need a good account of the metaphysics, epistemology, concept, and methodology of causality. In add-ition, there is a practical motivation of a good theory causation: we need a concept of cause which can be used, especially in scientific practice, to represent, explain, manipulate, and understand various phenomena in the world. We need such a theory of causation in order to make good sense of various phenomena around us. As Williamson (2006b, 269) indicates, causal beliefs are 'heuristic devices', which help reason about the world by means of making strategic decisions, predications, diagnoses, etc.

Woodward (2014, 694) argues that a functional approach has two important benefits by shedding light on the connections between causal concepts to elucidate the 'usefulness of causal thinking' and evaluate 'various strategies of causal cognition and other concepts', and leading to a focus on methodology to assess 'the usefulness of different causal concepts, and of procedures for relating causal claims to evidence', though he is implicit on the meaning of usefulness. In my functional approach to scientific progress (Shan 2019; 2022), I develop an account of usefulness: an exemplary practice (i.e. a particular way of problem-defining and problem-solving) is useful if and only if its way of defining and solving research problems is repeatable and provides a reliable framework for further investigation to solve problems and to generate novel research problems across different areas (or disciplines).

Thus, I propose that there is a fundamental desideratum that a good philosophical theory of causation should satisfy:

A good theory of causation ought to offer a good account of the nature of causation, and show how causation provides a conceptual tool contributing to a useful exemplary practice in order to represent, explain, manipulate, or understand various phenomena in the world.

Of course, we may never have such a comprehensive theory of causation. Nevertheless, I contend that any philosophical enquiry into causation without an examination of the practical issue is incomplete. In other words, a promising approach to causation ought to take all of the metaphysical, epistemological, conceptual, methodological, and practical issues into account.

References

Anjum, Rani Lill, and Stephen Mumford. 2010. 'A Powerful Theory of Causation.' In *The Metaphysics of Powers*, edited by Anna Marmodoro, 143–59. New York and London: Routledge.

Anjum, Rani Lill, and Stephen Mumford. 2018. *Causation in Science and the Methods of Scientific Discovery*. Oxford: Oxford University Press.

Anscombe, G. E. M. 1971. 'Causality and Determination.' In *Causation*, edited by Ernest Sosa and Michael Tooley, 88–104. Oxford: Oxford University Press.

Beach, Derek. 2021. 'Evidential Pluralism and Evidence of Mechanisms in the Social Sciences.' *Synthese* 199: 8899–919.

Bird, Alexander. 2010. 'Causation and the Manifestation of Powers.' In *The Metaphysics of Powers*, edited by Anna Marmodoro, 160–8. New York and London: Routledge.

Carroll, John W. 1990. 'The Humean Tradition.' *The Philosophical Review* 99 (2): 185–219.

Carroll, John W. 2009. 'Anti-Reductionism.' In *The Oxford Handbook of Causation*, edited by Helen Beebee, Christopher Hitchcock, and Peter Menzies, 279–98. Oxford: Oxford University Press.

Cartwright, Nancy. 2002. 'Causation: One Word, Many Things.' *Philosophy of Science* 71 (5): 805–19.

Cartwright, Nancy. 2007. 'Are RCTs the Gold Standard?' *BioSocieties* 2 (1): 11–20.

Cartwright, Nancy. 2010. 'What Are Randomised Controlled Trials Good For?' *Philosophical Studies* 147 (1): 59–70.

Chao, Hsiang-Ke, Szu-Ting Chen, and Roberta L. Millstein, eds. 2013. *Mechanism and Causality in Biology and Economics*. Dordrecht: Springer.

Clarke, Brendan, Donald Gillies, Phyllis Illari, Federica Russo, and Jon Williamson. 2014. 'Mechanisms and the Evidence Hierarchy.' *Topoi* 33: 339–60.

Collier, John. 1999. 'Causation Is the Transfer of Information.' In *Causation, Natural Laws, and Explanation*, edited by Howard Sankey, 215–63. Dordrecht: Kluwer Academic Publishers.

Crasnow, Sharon. 2011. 'Evidence for Use: Causal Pluralism and the Role of Case Studies in Political Science Research.' *Philosophy of the Social Sciences* 41 (1): 26–49.

Crasnow, Sharon. 2017. 'Process Tracing in Political Science: What's the Story?' *Studies in History and Philosophy of Science* 62 (1): 6–13.

Dowding, Keith. 2023. 'Process Tracing:Process Tracing:Causation and Levels of Analysis.' In *The Oxford Handbook of Philosophy of Political Science*, edited by Harold Kincaid and Jeroen van Bouwel, 328–42. Oxford: Oxford University Press.

Dowe, Phil. 2000. *Physical Causation*. Cambridge: Cambridge University Press.

Eagle, Antony. 2007. 'Pragmatic Causation.' In *Causation, Physics, and the Constitution of Reality: Russell's Republic Revisited*, edited by Huw Price, 156–90. Oxford: Clarendon Press.

Fales, Evan. 1990. *Causation and Universals*. London: Routledge.

Gillies, Donald. 2011. 'The Russo-Williamson Thesis and the Question of Whether Smoking Causes Heart Disease.' In *Causality in the Sciences*, edited by Phyllis McKay Illari, Federica Russo, and Jon Williamson, 110–25. Oxford: Oxford University Press.

Glennan, Stuart. 1996. 'Mechanisms and the Nature of Causation.' *Erkenntnis* 44: 49–71.

Glynn, Luke. 2011. 'A Probabilistic Analysis of Causation.' *British Journal for the Philosophy of Science* 62 (2): 343–92.

Graßhoff, Gerd, and Michael May. 2001. 'Causal Regularities.' In *Current Issues in Causation*, edited by Wolfgang Spohn, Marion Ledwig, and Michael Esfeld, 85–114. Paderborn: Mentis.

Hall, Ned. 2004. 'Two Concepts of Caustion.' In *Causation and Counterfactuals*, edited by John Collins, Ned Hall, and L. A. Paul, 225–76. Cambridge: The MIT Press.

Harré, Rom, and Edward H. Madden. 1975. *Causal Powers*. Oxford: Basil Blackwell.

Illari, Phyllis McKay, and Federica Russo. 2014. *Causality: Philosophical Theory Meets Scientific Practice*. Oxford: Oxford University Press.

Ingthorsson, Rögnvaldur Dadi. 2002. 'Causal Production as Interaction.' *Metaphysica* 3 (1): 87–119.

Ingthorsson, Rögnvaldur Dadi. 2021. *A Powerful Particulars View of Causation*. New York and London: Routledge.

Kistler, Max. 1998. 'Reducing Causality to Transmission.' *Erkenntnis* 48: 1–25.

Krauss, Alexander. 2021. 'Assessing the Overall Validity of Randomised Controlled Trials.' *International Studies in the Philosophy of Science* 34 (3): 159–82.

Lewis, David. 1973. 'Causation.' *Journal of Philosophy* 70: 556–67.

Lewis, David. 2000. 'Causation as Influence.' *Journal of Philosophy* 97: 182–97.

Machamer, Peter. 2004. 'Activities and Causation: The Metaphysics and Epistemology of Mechanisms.' *International Studies in the Philosophy of Science* 18 (1): 27–39.

Machamer, Peter, Lindley Darden, and Carl F. Craver. 2000. 'Thinking about Mechanisms.' *Philosophy of Science* 67 (1): 1–25.

Mackie, John Leslie. 1974. *The Cement of the Universe*. Oxford: Clarendon Press.

Maziarz, Mariusz. 2020. *The Philosophy of Causality in Economics: Causal Inferences and Policy Proposals*. London and New York: Routledge.

Maziarz, Mariusz. 2021. 'Resolving Empirical Controversies with Mechanistic Evidence.' *Synthese* 199: 9957–78.

Mumford, Stephen, and Rani Lill Anjum. 2011. *Getting Causes from Powers*. Oxford: Oxford University Press.

Northcott, Robert. 2012. 'How Necessary Are Randomized Controlled Trials?' In *Intervention and Reflection: Basic Issues in Medical Ethics*, edited by Ronald Munson, 9th ed., 187–91. Boston, MA: Wadsworth.

Norton, John D. 2003. 'Causation as Folk Science.' *Philosophers' Imprint* 3 (4): 1–22.

Reichenbach, Hans. 1956. *The Direction of Time*. Berkeley and Los Angeles, CA: University of California Press.

Reiss, Julian. 2012. 'Causation in the Sciences: An Inferentialist Account.' *Studies in History and Philosophy of Biological and Biomedical Sciences* 43: 769–77.

Reiss, Julian. 2015. *Causation, Evidence, and Inference*. New York: Routledge.

Reutlinger, Alexander. 2013. *A Theory of Causation in the Social and Biological Sciences*. New York: Palgrave Macmillan.

Runhardt, Rosa W. 2015. 'Evidence for Causal Mechanisms in Social Science: Recommendations from Woodward's Manipulability Theory of Causation.' *Philosophy of Science* 82 (5): 1296–307.

Runhardt, Rosa W. 2021. 'Evidential Pluralism and Epistemic Reliability in Political Science: Deciphering Contradictions between Process Tracing Methodologies.' *Philosophy of the Social Sciences* 51 (4): 425–42.

Russell, Bertrand. 1912. 'On the Notion of Cause.' *Proceedings of the Aristotelian Society* 13: 1–26.

Russo, Federica. 2009. *Causality and Causal Modelling in the Social Sciences.* Dordrecht: Springer.

Russo, Federica, and Jon Williamson. 2007. 'Interpreting Causality in the Health Sciences.' *International Studies in the Philosophy of Science* 21 (2): 157–70.

Salmon, Wesley C. 1980. 'Causality: Production and Propagation.' *PSA: Proceedings of the Biennial Meeting of the Philosophy of Science Association* 2: 49–69.

Shan, Yafeng. 2019. 'A New Functional Approach to Scientific Progress.' *Philosophy of Science* 86 (4): 739–58.

Shan, Yafeng. 2020. *Doing Integrated History and Philosophy of Science: A Case Study of the Origin of Genetics.* Cham: Springer.

Shan, Yafeng. 2022. 'The Functional Approach: Scientific Progress as Increased Usefulness.' In *New Philosophical Perspectives on Scientific Progress*, edited by Yafeng Shan, 46–61. New York: Routledge.

Shan, Yafeng, and Jon Williamson. 2021. 'Applying Evidential Pluralism to the Social Sciences.' *European Journal for Philosophy of Science* 11 (4). https://doi.org/10.1007/s13194-021-00415-z.

Shan, Yafeng, and Jon Williamson. 2023. *Evidential Pluralism in the Social Sciences.* London and New York: Routledge.

Suppes, Patrick. 1970. *A Probabilistic Theory of Causality.* Amsterdam: North-Holland Publishing Company.

Tooley, Michael. 1987. *Causation: A Realist Approach.* Oxford: Oxford University Press.

Tooley, Michael. 1990. 'Causation: Reductionism Versus Realism.' *Philosophy and Phenomenological Research* 50 (May): 215–36.

Wilde, Michael, and Veli-Pekka Parkkinen. 2019. 'Extrapolation and the Russo-Williamson Thesis.' *Synthese* 196 (8): 3251–62.

Williamson, Jon. 2005. *Bayesian Nets and Causality: Philosophical and Computational Foundations.* Oxford: Oxford University Press.

Williamson, Jon. 2006a. 'Causal Pluralism versus Epistemic Causality.' *Philosophica* 77: 66–96.

Williamson, Jon. 2006b. 'Dispositional versus Epistemic Causality.' *Minds and Machines* 16: 259–76.

Woodward, James. 2003. *Making Things Happen: A Theory of Causal Explanation.* Oxford: Oxford University Press.

Woodward, James. 2014. 'A Functional Account of Causation: Or, a Defence of the Legitimacy of Causal Thinking by Reference to the Only Standard That Matters— Usefulness (as Opposed to Metaphysics or Agreement with the Intuitive Judgment).' *Philosophy of Science* 81 (5): 691–713.

Worrall, John. 2010. 'Do We Need Some Large, Simple Randomized Trials in Medicine?' In *EPSA Philosophical Issues in the Sciences*, edited by Mauricio Suárez, Mauro Dorato, and Miklós Rédei, 289–301. Dordrecht: Springer.

Zwier, Karen R. 2017. 'Interventionist Causation in Thermodynamics.' *Philosophy of Science* 84 (5): 1303–15.

2

Causation

A Non-Reductionist, Theoretical-Term Analysis

Michael Tooley

1. Introduction

Reductionist approaches to causation have long dominated the field, and such approaches fall into two very different groups. On the one hand, there are approaches that accept the thesis of Humean supervenience advanced by David Lewis (1986a, ix–x)—which both captured and was inspired by David Hume's view that there cannot be any logical connections between spatially or temporally distinct existents. Within that group, the most important reductionist approaches are, first, ones that analyze causation in terms of regularities (Psillos 2009); secondly, approaches that employ counterfactuals either to analyze, or else provide an ontological reduction of, causation (Lewis 1973b, 1986b; Paul 2009; Paul and Hall 2013); and thirdly, approaches that analyze causation in terms of probabilistic relations found in the world between different types of events (Reichenbach 1956; Glynn 2011).

On the other hand, there are approaches that reject the Humean view that there cannot be logical connections between distinct existents, and that hold that among the ontologically basic things are what are variously referred to as powers, dispositions, propensities, and potencies (Harré and Madden 1975; Molnar 2003; Bird 2007; Mumford 2009; Demarest 2017). The contention is then that both causal relations and laws of nature are ontologically reducible to such entities.

My goal is to demonstrate the superiority of a non-reductionist analysis of the concept of causation to reductionist approaches of both of these general types.

But if the concept of causation cannot be analyzed in such ways, how can it be analyzed? Moreover, what is a non-reductionist analysis? The answer is that a non-reductionist analysis of any concept is the type of analysis that must be used for concepts of unobservable, theoretical entities—such as quarks and electromagnetic fields—so a non-reductionist analysis of the concept of causation treats causation as an unobservable, theoretical relation between events.

To set out a non-reductionist analysis of causation, one therefore needs a *theory* that involves certain *necessary truths* about the relation of causation. Here one could simply state the theory—which, as you will see, can be done rather quickly,

Michael Tooley, *Causation: A Non-Reductionist, Theoretical-Term Analysis* In: *Alternative Approaches to Causation: Beyond Difference-making and Mechanism*. Edited by: Yafeng Shan, Oxford University Press.
© Oxford University Press 2024. DOI: 10.1093/oso/9780192863485.003.0002

since it contains very few postulates—and then proceed to offer arguments in support of the theory. I think it is more illuminating, however, to indicate how one can *arrive* at the theory, as can be done by a simple thought experiment that immediately leads to the core idea in the non-reductionist analysis that I set out. That thought experiment is the focus of section 2.

Section 3 then describes the route that I followed from the thought experiment to the required postulates in the theory of causation, and section 4 contains the theoretical-term definition of the relation of causation.

Next, I consider the reasons that can be offered in support of the analysis—reasons that fall into two groups. First, there are highly desirable and very plausible properties of the analysis itself, and these are set out in section 5. Then there are objections to alternative accounts of causation, where; in section 6, I describe the three most general categories into which alternatives to the present analysis of the relation of causation fall, and then, in sections 7, 8, and 9, I set out objections that apply to the approaches that fall within each of those general types of accounts.

Section 10 then contains some very brief comments on related issues that I have not addressed, while section 11 provides a summary of my discussion.

2. A Thought Experiment and the Core Idea: Principles Linking Causation to Logical Probability

In approaching the task of setting out an analysis of the concept of causation, it is natural to attempt to find necessary truths involving that relation between events. It turns out, however, that the crucial necessary truths, rather than involving only the relation of causation, involve laws containing that relation. Accordingly, that is the route that will be followed here.

One could, then, simply set out an analysis of the concept of causal laws, and then show that, given that analysis, causal statements have the truth values that we normally take them as having. However, that approach does not seem satisfactory. For one thing, there might be logically possible worlds with very different causal connections from those found in our world, which one had thus failed to consider, and where the proposed analysis of the relation of causation would break down. Backward causation, for example, might be such a case, or causation across spatial or temporal gaps.

Let us turn, then, to a thought experiment that brings out the fundamental idea underlying my approach. That thought experiment involves considering events (or states of affairs) of two types, A and B, where A is some very simple type of event, while B is some very complex type of event. (A could consist of two neutrons colliding, or one's having a momentary experience of an instance of qualitative redness, while B could be an event involving the coming into existence of a planet whose state was qualitatively identical with that of our Earth at the beginning of the present millennium, and that was millions of light-years away in a certain direction from the event of type A.) In the absence of any evidence at all, one should surely view events of type A as much more likely than events of type B. Suppose, however, that an omniscient being informs one that events of type A occur when and only when corresponding events of type B occur, and, moreover, that this two-way relationship is entailed by some law of nature, which might be either a causal law or a non-causal law of simultaneous coexistence. Then one's initial probabilities certainly need to be adjusted. But how exactly should this be done? Should one assign a lower probability to events of type A, a higher probability to events of type B, or both? The answer, surely, is not at all clear, let alone precisely what new probability (or probabilities) should be chosen.

Suppose, however, that the omniscient being, wanting to be helpful, provides an additional piece of information—namely, that any event of type A is *causally sufficient and causally necessary* for a corresponding event of type B. How should one adjust the probabilities in that case? The answer, I suggest, is that one should adjust the probability that one assigns to events of type B, equating it with the probability one initially assigned to events of type A. Conversely, if one learns that events of type B are causally sufficient and causally necessary for events of type A, then the thing to do is to adjust the probability that one assigns to events of type A, equating it with the probability that one initially assigned to events of type B.

The relationships between *a priori* probabilities and *a posteriori* probabilities are very clear, then, in the case where events of one type are both causally sufficient and causally necessary for events of some other type. Thus, if we use 'CSN(A, B)' to say that it is a law that events of type A are both causally sufficient and causally necessary for corresponding events of type B, the following two very simple equations specify the relevant posterior probabilities:

$$(CSN_1) \ Prob(Ax/CSN(A, B)) = Prob(Ax)$$
$$(CSN_2) \ Prob(Bx/CSN(A, B)) = Prob(Ax)$$

In short, both the posterior probability of an event of type A and the posterior probability of an event of type B, relative to its being a law that events of type A are both causally sufficient and causally necessary for events of type B, are equal to the *a priori* probability of events of type A.

3. The Postulates for a Theoretical-Term Analysis
of the Concept of the Relation of Causation

I have suggested that the conclusion to be drawn from the preceding thought experiment is that while information to the effect that it is a law that events of type

A are both causally sufficient and causally necessary for events of type B does not give one any reason to think that the posterior probability of an event of type A, given that information, is different from the *a priori* logical probability of an event of type A, it does give one a reason for thinking that the posterior probability of an event of type B, given that information, is different from the *a priori* logical probability of an event of type B. So, can one use this idea to formulate an analysis of the concept of the relation of causation?

The answer is that one can, but to do that we need to shift from the cases where events of one type are both causally sufficient and causally necessary for events of some other type to cases where events of one type are causally sufficient, but need not be causally necessary for events of some other type, to arrive at the desired postulates that can be used to give a theoretical-term analysis of the concept of the relation of causation.

Consider, then, the two equations that emerged in the thought experiment:

$$(CSN_1)\ \text{Prob}(Ax/CSN(A, B)) = \text{Prob}(Ax)$$
$$(CSN_2)\ \text{Prob}(Bx/CSN(A, B)) = \text{Prob}(Ax)$$

How will these have to be changed when one shifts from the case where events of one type are both causally sufficient and causally necessary for some other type of event to the case where events of one type are *causally sufficient* (but may not be causally necessary) for some other type of event? In the case of the first equation, no change in *the general form* of the equation is needed, so the first postulate can be written as

$$\textbf{Prob}(\textbf{A}x/\textbf{L}(\textbf{C, A, B})) = \textbf{Prob}(\textbf{A}x) \tag{P_1}$$

—where the expression 'L(C, A, B)' says that it is a law that, for any x, x's being A causes x to be B—where x's being B could be a matter of there being some non-causal relation R such that there is a y that has some property P and that stands in relation R to x.

What about the posterior probability of an event of type B, given that it is a law that events of type A cause events of type B? Here the idea is that there are two ways that an event of type B can come about. One is that an event of type A occurs and that together with its being a law that, for any x, x's being A causes x to be B results in x's being B. The probability of this happening, given L(C, A, B), is just Prob(Ax/L(C, A, B)), which by postulate P_1 is equal to Prob(Ax).

The other way in which the state of affairs that consists of x's being B may come about is, so to speak, 'by accident' when Ax is not the case. The probability of this given that it is a law that L(C, A, B) is equal to the product of, first of all, the probability that Bx will be the case if both ~Ax and L(C, A, B) are the case and, secondly, the probability that $\sim Ax$ will be the case given $L(C, A, B)$. So the relevant probability that Bx will come about in that manner is $Pr(Bx/\sim Ax \& L(C, A, B)) \times Pr(\sim Ax/L(C, A, B))$.

Given that there are only these two possibilities for its being the case that x is B and that they are incompatible, given that one obtains only if Ax is the case, and the other obtains only if Ax is not the case, the probability that Bx is the case given that it is a law that $L(C, A, B)$ is the sum of those two probabilities. So we have

$$Prob(Bx/L(C, A, B)) = Prob(Ax/L(C, A, B)) + Pr(Bx/(\sim Ax \& L(C, A, B)) \times Pr(\sim Ax/L(C, A, B))) \tag{1}$$

It then follows from (1) given postulate (P_1) that

$$Prob(Bx/L(C, A, B)) = Prob(Ax) + Pr(Bx/(\sim Ax \& L(C, A, B))) \times Pr(\sim Ax/L(C, A, B)) \tag{2}$$

Next, there is the following theorem of conditional additivity from the formal theory of probability, which can be proved as follows (Hacking 2001, 61).

Let $Pr(E) > 0$. If A and B are mutually exclusive, then

$$Pr[(A \lor B)/E] = Pr[(A \lor B) \& E]/Pr(E) = Pr[(A \& E) \lor (B \& E)]/Pr(E) = Pr[(A \& E)/Pr(E)] + Pr[(A \& E)/Pr(E)] = Pr(A/E) + Pr(B/E)$$

So we have
Additivity
If $Pr(E) > 0$, then if A and B are mutually exclusive

$$Pr((A \lor B)/E) = Pr(A/E) + Pr(B/E)$$

This gives one that

$$Pr(Ax/L(C, A, B)) + Pr(\sim Ax/L(C, A, B)) = Pr[(Ax \lor \sim Ax)/L(C, A, B)] = 1 \tag{3}$$

According to postulate (P_1), however, one has that

$$Prob(Ax/L(C, A, B)) = Prob(Ax) \tag{4}$$

So (3) and (4) then give one:

$$Pr(Ax) + Pr(\sim Ax/L(C, A, B)) = 1 \tag{5}$$

This together with

$$Pr(Ax) + Pr(\sim Ax) = 1 \tag{6}$$

then entails

$$Pr(\sim Ax) = Pr(\sim Ax/L(C, A, B)) \tag{7}$$

Proposition (7) together with proposition (2) then entails

$$Prob(Bx/L(C, A, B)) = Prob(Ax) + Pr(Bx/(\sim Ax \, \& \, L(C, A, B))$$
$$\times \, Pr(\sim Ax) \tag{8}$$

Proposition (8) could be taken as a postulate, but the following line of thought suggests a different possibility. How could the existence of the law L(C, A, B) be relevant to the occurrence of a state of affairs of type B? If a state of affairs of type A occurs, the existence of the law is, of course, very relevant indeed since that law together with the existence of a state of affairs of type A entails that a state of affairs of type B will occur. By contrast, if a relevant state of affairs of type A does not exist, then it would seem that the existence of the law L(C, A, B) is not relevant at all: the probability of there being a state of affairs of type B will be no different from what it would be if the law L(C, A, B) did not exist. So it is natural to take the following proposition as a second postulate:

$$\textbf{Prob(B}x\textbf{/}\mathord{\sim}\textbf{A}x \, \& \, \textbf{L(C, A, B))} = \textbf{Prob(B}x\textbf{/}\mathord{\sim}\textbf{A}x\textbf{)} \tag{P$_2$}$$

Proposition (8) together with postulate (P$_2$) then entails the third and final postulate:

$$\textbf{Prob(B}x\textbf{/L(C, A, B))} = \textbf{Prob(A}x\textbf{)} + \textbf{Prob(B}x\textbf{/}\mathord{\sim}\textbf{A}x\textbf{)} \times \textbf{Prob(}\mathord{\sim}\textbf{A}x\textbf{)} \tag{P$_3$}$$

In the case of non-probabilistic laws, then, one has the following three principles:

$$\textbf{Prob(A}x\textbf{/L(C, A, B))} = \textbf{Prob(A}x\textbf{)} \tag{P$_1$}$$
$$\textbf{Prob(B}x\textbf{/}\mathord{\sim}\textbf{A}x \, \& \, \textbf{L(C, A, B))} = \textbf{Prob(B}x\textbf{/}\mathord{\sim}\textbf{A}x\textbf{)} \tag{P$_2$}$$
$$\textbf{Prob(B}x\textbf{/L(C, A, B))} = \textbf{Prob(A}x\textbf{)} + \textbf{Prob(B}x\textbf{/}\mathord{\sim}\textbf{A}x\textbf{)} \times \textbf{Prob(}\mathord{\sim}\textbf{A}x\textbf{)} \tag{P$_3$}$$

4. Defining the Relation of Causation

How does one move from these postulates to an analysis of the concept of the relation of causation? The starting point is with the idea that postulates (P$_1$)

through (P_3) serve to *define implicitly* the relation of causation, and the task is then to move from that to an explicit definition. How can that be done? The answer involves the solution to a problem that had existed at least since the time of the development of the atomic theory of matter by the British chemist John Dalton in 1803, and that became more pressing with the discovery of electrons by J. J. Thomson in 1904, and of protons by Ernest Rutherford in 1919. For how could one ever form those concepts, given that atoms were too small to be observed, and electrons and protons even more so?

A breakthrough occurred, however, in 1929, when a British philosopher, Frank Plumpton Ramsey, discovered a key idea, which he set out in his paper 'Theories.' It was not, however, a complete solution. Moreover, Ramsey did not himself accept a realist view of unobservable entities, instead favoring an instrumentalist view, according to which such theories, rather than referring to unobservable entities, instead serve merely as a way of moving from one set of propositions about *observable* entities to another set.

Over twenty years later, however, R. B. Braithwaite, in his book *Scientific Explanation* (1953, 79), mentioned, almost in passing, that Ramsey's idea could be used to set out non-reductionist analyses of concepts of unobservable entities. Very detailed and careful accounts were then set out, first by R. M. Martin in his paper 'On Theoretical Constructs and Ramsey Constants' (1966)—although Martin's account does not offer a robustly realist view of theoretical entities— and then a bit later, in 1970, David Lewis, in his paper 'How to Define Theoretical Terms,' described a way of defining theoretical terms that enables one to take a non-reductionist view of theories containing terms referring to unobservable entities, properties, and relations.

Let me now describe how, given the Ramsey/Lewis approach to the definition of theoretical terms, one can set out a non-reductionist analysis of the concept of the relation of causation. First, form the conjunction of the three postulates (P_1) through (P_3):

$$[\text{Prob}(Ax/L(C, A, B)) = \text{Prob}(Ax)] \ \&$$
$$[\text{Prob}(Bx/{\sim}Ax \ \& \ L(C, A, B)) = \text{Prob}(Bx/{\sim}Ax)] \ \&$$
$$[\text{Prob}(Bx/L(C, A, B)) = \text{Prob}(Ax) + \text{Prob}(Bx/{\sim}Ax) \times \text{Prob}({\sim}Ax)]$$

Next, replace the three predicates 'C', 'A', and 'B' with variables ranging over properties and relations. If we use 'P', 'Q', and 'R' for those variables, we have

$$[\text{Prob}(Qx/L(P, Q, R)) = \text{Prob}(Qx)] \ \&$$
$$[\text{Prob}(Rx/{\sim}Qx \ \& \ L(P, Q, R)) = \text{Prob}(Rx/{\sim}Qx)] \ \&$$
$$[\text{Prob}(Rx/L(P, Q, R)) = \text{Prob}(Qx) + \text{Prob}(Rx/{\sim}Qx) \times \text{Prob}({\sim}Qx)]$$

Then, since it is only 'C' that one wants to define, one affixes two universal quantifiers to the front of the resulting open sentence containing the variables that one has put in place of 'A' and 'B', so that one has an open sentence with only the one free variable—namely, 'P'—the variable that was used to replace all occurrences of 'C':

$$(\forall Q)(\forall R) \{[Prob(Qx/L(P, Q, R)) = Prob(Qx)] \ \&$$
$$[Prob(Rx/\sim Qx \ \& \ L(P, Q, R)) = Prob(Rx/\sim Qx)] \ \&$$
$$[Prob(Rx/L(P, Q, R)) = Prob(Qx) + Prob(Rx/\sim Qx) \times Prob(\sim Qx)]\}$$

The relation of causation is then defined as that unique relation between states of affairs that satisfies the open sentence in question.

5. Some Highly Desirable Aspects of This Account of the Relation of Causation

Let me now describe some highly desirable properties of this analysis.

5.1 A First Desirable Feature: The Simplicity of the Analysis

First of all, the theoretical-term, non-reductionist analysis of the concept of the relation of causation is extremely simple, and in this respect, it contrasts very sharply with, for example, the two most popular types of reductionist approaches—namely, probabilistic approaches and counterfactual approaches.

As regards the former, the most circumspect development of a reductionist probabilistic approach to causation is that set out by Luke Glynn in his 2011 article 'A Probabilistic Analysis of Causation.' Glynn's route to his analysis involves introducing the following concepts, some of which are themselves quite complicated: (i) the idea of a set S of variables being *a revealer of positive relevance* of C to E; (ii) the idea of a *positive component* effect; (iii) the idea of being a *positive relevance eliminator set* for C and E; (iv) the idea of being a *stable positive relevance eliminator set* for C and E; (v) the idea of C's being an *unneutralized positive component effect* upon E; and (vi) the idea of C's being a *positive token cause* of E.

Does one, in the end, at least have a satisfactory analysis of the concept of causation? The answer is that one does not since, first of all, Glynn has to appeal to the relation of temporal priority to provide a direction for causation; secondly, Glynn's account entails that if C is a *direct* cause of E, then C must raise the probability of E, and a very slight variation on an argument advanced by Wesley Salmon in his 1980 article 'Probabilistic Causality' shows that this is false; and thirdly, Glynn is exposed to the 'underdetermined' and 'temporally inverted worlds' objections to be set out in section 7.

How does the present account compare with counterfactual approaches? The initial counterfactual account set out by David Lewis (1973b) was based upon a Stalnaker (1968)/Lewis (1973a) style account of counterfactuals, to which a very strong objection was advanced in reviews by Jonathan Bennett (1974) and Kit Fine (1975). Lewis, in 'Counterfactual Dependence and Time's Arrow' (1979, 472; 1986; 47–8) attempted to show that, given the following account of the weight to be assigned to factors that enter into judgments of similarity, one could escape the objection that had been advanced by Bennett and Fine:

(1) It is of the first importance to avoid big, widespread, diverse violations of law.
(2) It is of the second importance to maximize the spatio-temporal region throughout which perfect match of particular fact prevails.
(3) It is of the third importance to avoid even small, localized, simple violations of law.
(4) It is of little or no importance to secure approximate similarity of particular fact, even in matters that concern us greatly.

I think it doubtful that Lewis's intuitions here would be widely shared. In particular, might not one think that a complete match with regard to all future events counted for more than big, widespread, and diverse violations of laws of nature *at a single time*? In any case, I showed (Tooley 2003) that even if one accepts (1) through (4), the objection advanced by Bennett and Fine can be revised so that it is successful.

As a result, Lewis's formulation of a counterfactual approach to causation has been more or less completely abandoned in favor of other counterfactual approaches, with one of the most recent being that advanced by L. A. Paul and Ned Hall in their book, *Causation: A User's Guide*, where they abandon the attempt to offer *an analysis* of the concept of causation in favor of giving a counterfactual account of what causation is in the actual world. So how does that more modest program fare as regards simplicity? The answer is that a chapter that deals only with cases of redundant causation requires more than 100 pages!

5.2 A Second Desirable Feature: Causation and Verbs, Transitive and Intransitive, Dealing with Simple Actions

The simplicity of the non-reductionist analysis of causation is, moreover, a desirable feature in another way. For consider the fact that every known language surely contains an enormous number of verbs that involve the concept of the relation of causation—not only transitive verbs where one performs some action to affect some object, such as raising a glass, but also intransitive verbs, such as sitting down, or going for a walk, where one's physical movements result from intentions that one had. If causation were as complicated as it is on reductionist accounts, would it be at all plausible not only that the number of terms involving that concept is enormous today, but also that such terms must have been present from the time that humans acquired the capacity for language?

5.3 A Third Desirable Feature: Two Theorems Concerning Increases in Probabilities

Another desirable intrinsic feature of the above account is that one can derive propositions, for both probabilistic and non-probabilistic laws, stating that causes raise the probabilities of their effects. In the case of non-probabilistic laws, the proof is based on the third of the three postulates, namely

$$\text{Prob}(Bx/L(C, A, B)) = \text{Prob}(Ax) + \text{Prob}(Bx/\sim Ax) \times \text{Prob}(\sim Ax) \qquad (\text{P}_3)$$

This postulate enables one to offer an extremely simple proof of the crucial theorem concerning the relation between causation and probabilities. It involves comparing (P_3) with an instance of the following theorem in the formal theory of probability:

$$\text{Prob}(p) = \text{Prob}(p/q) \times \text{Prob}(q) + \text{Prob}(p/\sim q) \times \text{Prob}(\sim q) \qquad (1)$$

the relevant instance of which is

$$\text{Prob}(Bx) = \text{Prob}(Bx/Ax) \times \text{Prob}(Ax) + \text{Prob}(Bx/\sim Ax) \times \text{Prob}(\sim Ax) \qquad (2)$$

Consider now subtracting the left-hand side of equation (2) from the left-hand side of (P_3), and the right-hand side of (2) from the right-hand side of (P_3):

$$\text{Prob}(Bx/L(C, A, B)) = \text{Prob}(Ax) + \text{Prob}(Bx/\sim Ax) \times \text{Prob}(\sim Ax) \qquad (\text{P}_3)$$
$$\text{Prob}(Bx) = \text{Prob}(Bx/Ax) \times \text{Prob}(Ax) + \text{Prob}(Bx/\sim Ax) \times \text{Prob}(\sim Ax) \qquad (2)$$

As one can see, the result is

$$\text{Prob}(Bx/L(C, A, B)) - \text{Prob}(Bx) = \text{Prob}(Ax) - \text{Prob}(Bx/Ax) \times \text{Prob}(Ax)$$

which can be rewritten as

$$\text{Prob}(Bx/L(C, A, B)) - \text{Prob}(Bx) = \text{Prob}(Ax)(1 - \text{Prob}(Bx/Ax)) \qquad (3)$$

Then, since it cannot be the case that Prob(Ax) is equal to zero unless Ax is a contradiction, and it cannot be the case that Prob(Bx/Ax) is equal to 1 unless either Ax entails Bx, or else Bx is a necessary truth, one has that

$$Prob(Ax)(1 - Prob(Bx/Ax)) > 0 \qquad (4)$$

It then follows from (3) and (4) that

$$Prob(Bx/L(C, A, B)) - Prob(Bx) > 0 \qquad (5)$$

Thus we have

The First Increase of Probability Theorem

The logical probability that Bx is the case given that it is a law that events of type A cause events of type B is greater than the *a priori* probability of Bx.

Moreover, a similar theorem can be shown to hold in the case of probabilistic causal laws. Thus, if 'M(C, A, B, k)' is the proposition that it is a law that the probability that a state of affairs of type A causes a state of affairs of type B is equal to k, the relevant postulates turn out to be

$$Prob(Ax/M(C, A, B, k)) = Prob(Ax) \qquad (Q_1)$$
$$Prob(Bx/(\sim Ax \ \& \ M(C, A, B, k))) = Prob(Bx/\sim Ax) \qquad (Q_2)$$
$$Prob(Bx/M(C, S, B, k)) = (k \times Prob(Ax)) + ((1 - k) \times Prob(Bx/Ax)$$
$$\times Prob(Ax)) + (Prob(Bx/\sim Ax) \times Prob(\sim Ax)) \qquad (Q_3)$$

These postulates can then be shown to entail

The Second Increase of Probability Theorem

The logical probability that Bx is the case given that it is a law that states of affairs of type A cause states of affairs of type B with probability k is greater than the *a priori* probability of Bx.

In contrast, as mentioned above, a version of an argument advanced by Wesley Salmon shows that, given a reductionist, probabilistic analysis of causation, it can be the case that an event of type A is the direct cause of an event of type B, even though the occurrence of the event of type A made the occurrence of an event of type B less probable than it would have been if the event of type A had not occurred, since if an event of type A had not occurred, a different type of event would have occurred that was more likely to give rise to an event of type B.

5.4 A Basis for a Causal Analysis of the Earlier-Than Relation

We have seen that the postulates involved in the analysis of causation entail that causes raise the probabilities of their effects, whereas the probabilities of causes are not, in the same way, a function of their effects, and this asymmetry in these probabilistic relations provides the basis for a causal analysis of the earlier-than relation. This is a highly desirable outcome, since as we shall see below, in section 6.2, other accounts of the relation of temporal priority are open to strong objections.

6. The Alternatives to a Non-Reductionist Analysis of Causation

What types of alternatives are there to a theoretical concept, non-reductionist analysis of the concept of the relation of causation? The answer is that there are three general types of alternatives. First, there are Humean reductionist views that hold that there are no logical connections between distinct existents, where the most important competing theories are counterfactual theories, regularity theories, relative frequency probabilistic accounts, agency and interventionist theories, mechanistic theories, and conserved qualities and continuous processes accounts.

Next, there are non-Humean reductionist views that hold that there can be some logical connections between distinct existents. Such views hold that causation can be reduced to what are known as powers, dispositional properties, or potencies, where the latter are viewed as irreducible, intrinsic properties.

Finally, there is the approach to the concept of causation that is the simplest of all—namely, the non-reductionist view according to which causation is a directly observable relation, so that the concept of causation is analytically basic.

7. Some Central Objections to Humean Reductionist Analyses

Objections to Humean reductionist views are of two sorts. On the one hand, there are objections to specific views. Thus, in the case of reductionist probabilistic analyses of the concept of causation, a central objection is that contrary to such views, a cause need not raise the probability of its effect, even in the case of a direct cause, while in the case of David Lewis's original counterfactual analysis, the crucial objection is that Lewis's possible worlds account of counterfactuals is unsound. In the case of agency and interventionist views, they suffer from circularity, while conserved quantity views are open to the central objection that there can be causal laws where no quantity is conserved, while continuous processes approaches to causation are open to the central objection that causation across spatial and temporal gaps is logically possible.

On the other hand, there are objections that tell against *all or virtually all* Humean reductionist accounts, of which perhaps the four most important types are (1) 'underdetermination' objections, (2) 'direction of causation and time' objections, (3) 'simple worlds' objections, and (4) 'temporally inverted worlds' objections.

7.1 'Underdetermination' Objections

This type of objection can be set out in a variety of ways. One way appeals to the idea that uncaused events are logically possible and combines that with the idea of probabilistic causal laws to argue for the conclusion that there could be events where an event of a given type either was caused in accordance with some probabilistic causal law or else was simply an uncaused event. Thus there would be nothing in the world that made one of those things the case rather than the other.

A second way of setting out an 'underdetermination' objection involves considering a logically possible world in which there are two probabilistic causal laws, one linking events of type P as causes to events of type R as effects, and the other linking events of type Q as causes to events of type R as effects, and the idea is in cases where events of types P and Q occur together, there will be nothing to distinguish among (a) cases where the event of type P was the cause, (b) cases where the event of type Q was the cause, and (c) cases where there was causal overdetermination.

A third way of setting out the argument involves, instead, the claim that *non-probabilistic*, but *indeterministic* causal laws of the following form are possible: the existence of an event involving a specific property P causes the existence of an event that involves either or both of two properties, Q and R.

Consider, then, a world with two non-probabilistic, but indeterministic causal laws of that sort:

(1) It is a law that, for any object x, and any time t, x's acquiring property P at time t causes x either to acquire property Q at time $(t + \triangle t)$, or to acquire property R at time $(t + \triangle t)$, or to acquire both property Q and property R at time $(t + \triangle t)$.

(2) It is a law that, for any object x, and any time t, x's acquiring property S at time t causes x either to acquire property Q at time $(t + \triangle t)$, or to acquire property R at time $(t + \triangle t)$, or to acquire both property Q and property R at time $(t + \triangle t)$.

Suppose, finally, that some object a acquires both property P and also property S at time t, and that a then acquires both property Q and property R at time $(t + \triangle t)$, so that the situation is as follows:

Time t	Time $(t + \triangle t)$
Pa and Sa	Qa and Ra

Assume, finally, that there are no other factors that are causally relevant to a's coming to have property Q, or to a's coming to have property R, and that the world is one where uncaused states of affairs do not exist. Then, given (1) and (2), there are the following seven possibilities concerning the relevant causal relations:

Possibility 1: Pa caused Qa, and Sa caused Ra

Possibility 2: Pa caused Ra, and Sa caused Qa

Possibility 3: Pa caused Qa and Ra, and Sa caused Ra

Possibility 4: Pa caused Qa and Ra, and Sa caused Qa

Possibility 5: Pa caused Qa, and Sa caused Qa and Ra

Possibility 6: Pa caused Ra, and Sa caused Qa and Ra

Possibility 7: Pa caused Qa and Ra, and Sa caused Qa and Ra

All seven of these possibilities are compatible with all of the non-causal states of affairs, and with the two indeterministic causal laws. Accordingly, it is possible for there to be causal relations that do not logically supervene upon the combination of (1) the totality of states of affairs involving only non-causal properties and non-causal relations, (2) all of the laws of nature, both causal and non-causal, and (3) the general direction of causation in the world in question.

Notice, too, that this argument can easily be modified to provide an objection to non-Humean reductionist approaches that attempt to analyze the concept of causation in terms of non-reductionist concepts of intrinsic powers, dispositions, and the like.

7.2 The 'Direction of Causation and the Direction of Time' Objection

What is it that provides the basis for the relation of causation having a direction in reductionist theories? Many contemporary reductionist approaches incorporate the concept of temporal priority into their accounts. This is the case, for example, with Luke Glynn's probabilistic analysis (2011), and with the modified counterfactual account of what causation is in the actual world that is set out by L. A. Paul and Ned Hall (2013).

Why is this problematic? The answer is that one then needs to give an account of the concept of temporal priority, other than a causal account. So what are the options? There would seem to be only three. One appeals to certain patterns found in time, such as that of outgoing waves in water when a stone strikes the surface of a pond, or irreversible processes, such as the shattering of a glass, or the direction of increases in entropy. A second possibility is a tensed account, where the earlier-than relation is analyzed in terms of tensed notions such as those of past, present, and future. A third possibility involves claiming that one can directly observe one event's being earlier than another, so that temporal priority is an analytically basic concept.

What are the problems with such approaches to the concept of temporal priority? As regards the first, according to which the direction of time is to be analyzed in terms of certain patterns of events in time, we shall see in sections 7.3 and 7.4 that it is exposed to two types of objections that tell against virtually all Humean reductionist approaches to causation—the 'simple worlds' objection and the 'temporally inverted worlds' objection.

What about the second idea, according to which the earlier-than relation can be analyzed in terms of the tensed concepts of the past, the present, and the future? The basic objection is as follows. Firstly, any tensed analysis of the concept of temporal priority requires not just the concept of the present, but also the concepts of the past and the future. Secondly, the concept of the future cannot be analytically basic, since one is not immediately aware of, nor does one directly perceive, any future events. Thirdly, when one asks how the concept of the future is best analyzed, the natural answer is that the future is simply the concept of what is later than the present. Given, however, that the later-than concept is simply the inverse of the earlier-than concept, any analysis of temporal priority in terms of the tensed concepts of pastness, presentness, and futurity is therefore necessarily implicitly circular.

Finally, what about the third possibility—defended by Nathan Oaklander—that the relation of temporal priority is 'primitive and unanalyzable' (2004, 24)? This view is open to at least two serious objections. First of all, temporal priority has certain formal properties. It seems, for example, to be a necessary truth that temporal priority is a transitive relation. If the concept of temporal priority is analyzable, one may be able to show that it is an *analytic* truth that temporal priority is transitive, whereas if one holds instead that the concept of temporal priority is not analyzable, one is forced to view it as a synthetic *a priori* truth that temporal priority is a transitive relation.

Secondly, it seems plausible, first of all, that, aside from logical and mathematical concepts, all analytically basic concepts pick out properties or relations that can be immediately perceived, or of which one can be immediately aware, and secondly, that one's perception of one event as being earlier than another event always involves memories of immediately preceding perceptions, and that in being aware of a memory *as a memory* one has a belief that the mental state in question is causally related to the earlier state that is being remembered. If this is right, then one's awareness that one event is earlier than another is a case of *mediate* perception. Accordingly, the concept of temporal priority cannot be analytically basic.

The conclusion, in short, is that, unless the direction of time can be analyzed in terms of patterns of events in time, there does not appear to be any viable alternative to a causal analysis of the concept of temporal priority. The next two objections will show, however, that the direction of time cannot be analyzed in terms of patterns of events in time, and thus that there is no viable alternative to a causal analysis of the concept of temporal priority.

7.3 'Simple World' Objections

Our world is a complex one, with different features that might be viewed as providing the basis of a reductionist account of the direction of causation. First of all, it is a world where the direction of increase in the physical quantity known as entropy is the same in the vast majority of isolated or quasi-isolated systems (Reichenbach 1956, 117–43; Grünbaum 1973, 254–64).

Secondly, the temporal direction in which order is propagated—for example, by the circular waves resulting when a stone strikes a pond or by the spherical wavefronts associated with a point source of light—is invariably the same: the waves always move outward, never inward (Popper 1956, 538).

Thirdly, consider the causal forks involved when two events have either a common cause or a common effect. A fork may be described as open if it does not involve both a common cause and a common effect and closed if it involves both a common cause and a common effect. Given that distinction, it has then been claimed that it is a fact about our world that all or virtually all open forks are open in the same direction—namely, toward the future (Reichenbach 1956, 161–3; Salmon 1978, 696).

Can such features provide a satisfactory account of the direction of causation? One type of objection involves possible worlds that are much simpler than our own, and here are two very simple possible worlds.

(1) **Two Endlessly Rotating Spheres.** This is a world that contains no force fields, either gravitational or electromagnetic, and no material objects except for two spheres, connected by a rod, that rotate endlessly about the center of gravity of the system, in circular trajectories, following the laws of motion of Newtonian physics. This is a very simple world that by definition involves causal relations since the velocities of the two spheres are constantly changing in direction due to the forces that are exerted on them by the rod connecting them. But it is also a world that is utterly devoid of changes in entropy, of propagation of order, and of open forks. So there is no hope of basing an account of the direction of causation upon any of those features.

(2) **A Single-Particle World.** Here, however, is an even simpler world. Like the first, it contains no force fields at all, either gravitational or electromagnetic, but it is also a world devoid of *all* forces. All that one has is a single uncharged particle that persists through time, and whose properties never change.

Unlike the world with the two endlessly rotating spheres, the description of the single-particle world does not refer to any forces, and so there is no *immediate* reference to causal relations. It can be argued, however, that persistence through time presupposes causal relations, so that even the very simple, single-particle world involves causal relations.

Since the direction of causation in such simple worlds cannot be analyzed in terms of patterns of events in time, the only alternative open to a reductionist analysis of causation is to base the direction of causation upon the direction of time itself, but we have already seen, in section 7.2, that that alternative is ruled out. Consequently, Humean reductionist approaches to causation cannot account for causal relations in simple universes.

7.4 The Temporally Inverted Worlds Objection

Let us now move on to the fourth and final objection. Here the claim is that there are extremely complex possible worlds—as complex as the actual world—that contain causal processes for which Humean reductionist accounts, rather than generating no answer at all concerning the direction of those causal processes, yield wrong answers.

It is the year 4004 BC, and a Laplacian-style deity with a sense of humor—not a virtue normally attributed to deities—is about to create a world rather similar to ours, but one where Newtonian physics is true. Having selected the year AD 3000 as a good time for the world to be destroyed, the deity then works out what the world will be like at that final point, down to the last detail. He then creates two spatially unrelated worlds: the one just mentioned, together with another whose initial state is what I shall refer to as a 'flipped-over' version of the state of the first world at the final moment of its existence, where what this means is that the initial state of the second world is exactly like the final state of the first world, except that the velocities of the particles in the one world are exactly opposite to those of the corresponding particles in the other world.

Consider, now, any two complete temporal slices of the first world—A and B—where A is earlier than B. Since the worlds are Newtonian ones, and since the laws of Newtonian physics are invariant with respect to time reversal, the second world, which starts off from the reversed, AD 3000-type state, will go through corresponding states, B* and A*, where these are flipped-over versions of B and A respectively, and where B* is earlier than A*. So while one world goes from a

4004 BC, Garden of Eden state to an AD 3000, pre-annihilation state, the other world will move from a reversed version of the pre-annihilation type of state to a reversed version of the Garden of Eden type of state.

In the first world, the direction of causation will coincide with such things as the direction of increase in entropy, the direction of the propagation of order in non-entropically irreversible processes—such as the outgoing concentric waves produced in a pond by an object hitting the surface—and the direction defined by all (or most) open forks. But in the second world, where the direction of causation runs from the initial state created by the deity—that is, the flipped-over, AD 3000 type of state—through to the flipped-over, 4004 BC type of state, the direction in which entropy increases, the direction in which order is propagated, and the direction defined by open forks will all be in the opposite direction—that is, from temporally later states to temporally earlier ones. So if any of the latter patterns in events were used to define the direction of causation, doing so would generate the wrong result for the direction of causation in the case of the second world.

How might one attempt to escape this objection? Two possibilities come to mind. First of all, one might try to make use of the fact that if, in the case of the temporally flipped-over world, one traces things back in the direction defined by the open forks—which is, as a matter of fact, in the direction of earlier and earlier states of affairs—one eventually gets back to an open fork pointing in the opposite direction: namely, the fork involved in the creation by the omnipotent deity of the two spatially isolated worlds.

In the case of this response, one could very well ask what account of the direction of causation can be formulated according to which that one open fork will outweigh all the subsequent open forks that point in the opposite direction. But one can also reformulate the case to avoid any initial creative open fork. This could be done, for example, by having each of the two spatially separated worlds created by different deities. Alternatively, one could argue that if two such worlds could be created by a deity, they could also have simply popped into existence uncaused.

A second possible response to this 'temporally inverted' worlds objection is, of course, to hold that the direction of causation is defined not by patterns of events in time, but by the direction of time itself. But, as was argued in the case of the 'simple universes' argument, such a response is unsatisfactory.

8. The Fundamental Objection to a Non-Humean Reductionist Analysis

Non-Humean reductionist analyses of causation involve the idea of *irreducible, intrinsic powers*, and the fundamental objection to such an approach is that the idea of irreducible, intrinsic powers is incoherent. To set out that objection, one needs, first of all, the ideas of intrinsic properties and intrinsic relations, and then, secondly, the idea of an intrinsic state of affairs.

How are *properties* to be understood here? First, the idea of properties (and, similarly, relations) must be understood in such a way that Nelson Goodman's (1955) concepts of grue and bleen do not pick out properties. Similarly, negative properties and disjunctive properties, so-called, for example, must also be ruled out. Consequently, there is no one-to-one correspondence between concepts that do not involve names of particulars, on the one hand, and properties and relations on the other. Nor are sets of ordered pairs of individuals and possible worlds, though they may correspond to properties, themselves properties. To treat them as such would mean that there would be no non-circular way of defining the idea of a possible world.

Next, what are *intrinsic* properties? The basic idea that one wants to capture here is that of properties that are in no way dependent upon the existence of states of affairs that are external to the object that has the property. A natural, first attempt at capturing this idea might be as follows:

P is an intrinsic property = def.

P is a property and an entity x can have property P even if no particular outside of x exists.

This definition does not quite succeed, however, in capturing the intuitive notion of an intrinsic property. For consider the property of being the only electron that exists. An entity could have that property even if nothing else existed. But its having that property is dependent upon the *non-existence* of other electrons, and so the property is not really an intrinsic one.

Here, however, is a natural way of revising the definition that avoids that problem:

P is an intrinsic property = def.

P is a property and it is logically possible for an entity e to have property P regardless of what other particulars outside of e exist or fail to exist.

This idea then needs to be extended from the case of properties to that of relations. Thus, in the case of relations that involve only two relata—dyadic relations—one has the following:

R is an intrinsic dyadic relation = def.

R is a dyadic relation, and two entities e and f can stand in relation R regardless of what other particulars outside of e and f exist or fail to exist.

This is then easily generalized to the case of a relation holding among n things:

R is an intrinsic n-adic relation = def.

R is an n-adic relation, and n entities $e_1, e_2, \ldots e_n$ can stand in relation R to one another regardless of what other particulars outside of those n entities exist or fail to exist.

Given the ideas of intrinsic properties and intrinsic relations, one can then define the idea of an intrinsic *state of affairs*:

S is an intrinsic state of affairs = def.

S is a state of affairs, and every property and relation that is a constituent of S is either an intrinsic property of some particular involved in S or an intrinsic relation between particulars involved in S.

Next, suppose that state of affairs S is an intrinsic state of affairs and that all of the particulars involved in S are located within some *spatial region* R, while T is an intrinsic state of affairs involving at least one particular, x, that is not located within spatial region R, and thus which is not identical with any particular that is a constituent of state of affairs S. Consequently S, being an intrinsic state of affairs, could exist even if x did not exist, whereas state of affairs T could not, so the existence of state of affairs S cannot entail the existence of state of affairs T. So we have

First Conclusion Concerning Intrinsic States of Affairs

No intrinsic state of affairs can entail the existence of a *spatially distinct* intrinsic state of affairs.

Similarly, suppose that state of affairs S is an intrinsic state of affairs and that all of the particulars involved in S are located *at time t*, while T is an intrinsic state of affairs involving at least one particular, x, that is not located at time t, and thus which is not identical with any individual that is a constituent of state of affairs S. Then S, being an intrinsic state of affairs, could exist even if x did not exist, whereas state of affairs T could not, so the existence of state of affairs S cannot entail the existence of state of affairs T. Thus we also have

Second Conclusion Concerning Intrinsic States of Affairs

No intrinsic state of affairs can entail the existence of a *temporally distinct* intrinsic state of affairs.

Consider, first of all, *conserved quantities*, such as electric charge. Given an ontology of irreducible powers, unaccompanied by governing laws of nature, then wherever an electric charge is present at some time there must be a non-conditional irreducible power in virtue of which a corresponding electric charge is present at some location or other at later times. But if irreducible powers are *intrinsic* properties of objects, then an instantaneous particular's having the non-conditional power in question at time t is an intrinsic state of affairs—call it S—as is the state of affairs—call it T—that consists of there being some instantaneous particular that has the same non-conditional power at a later time t^*, and these intrinsic states of affairs are temporally distinct. Since, as we have just seen, no intrinsic state of affairs can entail the existence of a temporally distinct intrinsic state of affairs, it must be logically possible for S to exist without T's existing. But if the charge conservation property is an irreducible power, then the existence of S must, given the definition of the irreducible power in question, entail the existence of T. The idea of the non-conditional, irreducible, intrinsic power property in question therefore leads to a contradiction.

The same is true of *non-conditional*, irreducible, intrinsic power properties that do not involve conserved quantities. For example, in the case of the relation between a charged particle and the corresponding electric field, the existence of a charged particle at a given location at time t causally gives rise to an electric field at other locations whose strength and time of existence t^* depend upon the strength of the charge and the distance between the location of the charge and the other point. If the non-conditional, irreducible power in question is an intrinsic property, then a particle's having that property at time t is an intrinsic state of affairs, as is the existence of an electric field of a certain strength at a certain location at a later time t^*. By the definition of the non-conditional, irreducible power in question, the existence of the former state of affairs must entail the existence of the latter state of affairs, but by the definition of an intrinsic state of affairs, and the fact that the two states of affairs in question are both temporally and spatially distinct, the existence of the former state of affairs cannot entail the existence of the latter. So the concept of a non-conditional, irreducible, intrinsic power is once again seen to be self-contradictory.

Finally, the situation is no different in the case of *conditional*, irreducible, intrinsic powers. Consider, for example, the case of water-solubility. Given some fundamental particles of certain types in relevant spatial relations, one has a piece of salt, while given some other fundamental particles in relevant spatial relations, one has a group of water molecules. For the piece of salt to be in water at a given time is then simply a matter of the relevant spatial relation between the two groups of particles at that time. But then, according to an ontology of irreducible, intrinsic powers, any fundamental particle is what it is because of its irreducible powers, and since all of the powers involved are intrinsic properties, it follows that a piece of salt's being in water at a given time t is an intrinsic state of affairs. Similarly, a piece of salt being partially dissolved in water at a slightly later time t^* is also an intrinsic state of affairs. Since those two intrinsic states of affairs are in different temporal locations, it follows that the existence of the one cannot entail the existence of the other. But if water-solubility were an irreducible power, the piece of salt having that property at one time and being in water at that time would have to entail the salt's undergoing dissolving at some adjoining later time, however short. So the idea that water-solubility is an irreducible, conditional power leads to a contradiction.

There is, however, nothing special about the water-solubility case: the situation is exactly the same for any conditional, irreducible power property. The conclusion, accordingly, is that such properties are logically impossible.

The overall conclusion, then, is that a non-Humean reductionist attempt to analyze the concept of causation in terms of intrinsic irreducible powers or dispositions must be rejected, since the concept of irreducible, intrinsic power properties leads to a contradiction, and thus is incoherent.

9. A Refutation of the Simplest Non-Reductionist View

Direct causal realism is the view that (1) the concept of the relation of causation is analytically basic; (2) it is possible to be directly aware that two states of affairs stand in the relation of causation; and (3) the relation of causation is not reducible to non-causal properties and relations, nor to non-causal properties and relations together with causal laws, nor to anything else.

What is one to say about this approach to the concept of causation? First of all, the third claim is also part of the approach to causation that I have set out and defended, so, given my view, it is the first two claims that are crucial. Secondly, those first two claims are not, I would claim, unrelated, since it seems to me that the only way of defending the first claim is by appealing to the second claim.

Let us start, however, with the first claim. First of all, if one is going to claim that any concept is analytically basic, one needs to offer some criterion that a concept must satisfy if it is to be analytically basic. Secondly, in the case of descriptive concepts—as contrasted with logical and mathematical concepts—it is very plausible that a necessary condition of being a concept that is incapable of being analyzed is that the concept applies to states of affairs in virtue of properties or relations that one *can be directly aware of*. Thirdly, a property or relation cannot be such as one can be directly aware of unless, as David Hume and the British Empiricists maintained, that property or relation can be immediately given in experience. Fourthly, a property or relation cannot be *immediately* given in an experience E unless it would also be given in any experience E* that was *qualitatively identical* with E.

If the concept of the relation of causation is to be analytically basic, then, given that it is a descriptive concept, rather than a logical or mathematical one, the relation of causation must be immediately given in experience. Is this the case? The answer is that it is not. For given any experience E whatever—regardless of whether it involves a perception of external events, or an awareness of pressure upon one's body, or an introspective awareness of some mental occurrence, such as an act of willing, or a process of thinking—it is logically possible that appropriate, direct stimulation of the brain might produce an experience, E*, that was qualitatively identical with E, but that did not involve any causally related elements. So, for example, consider the experience of deciding to raise one's arm, followed immediately by an experience of one's arm's rising. Even if both experiences were *in themselves* veridical, there need not be any causal connection between one's deciding to raise one's arm and the rising of one's arm. It could be, for example, that a naughty deity with a sense of humor had severed the motor pathway leading from one's brain to muscles in one's arm, but had also directly stimulated the relevant muscle so that it contracted, raising one's arm, quite by accident, at the relevant time. Similarly, it might seem to one that one was engaging in a process of deductive reasoning, when, in fact, there was not any direct connection at all within the sequence of thoughts themselves—all of them being caused instead by one or more things outside of oneself, or by internal states that were completely independent of earlier thoughts. Causal relations cannot, therefore, be immediately given in experience in the sense that is required if the concept of causation is to be unanalyzable.

10. Logical Probability and the Epistemological Problem of Laws of Nature

There are two issues that I have not been able to address, since each would require an essay at least as long as this one. The first is connected with the non-reductionist analysis of the relation of causation that I have set out, and it concerns the idea of logical probability. The second is a general skeptical problem that arises for any account of causal laws.

As regards the first, many philosophers are skeptical about the idea of logical probability. Such skepticism, though understandable, is ill-advised, since no attempt to justify inductive inferences that does not involve a theory of logical probability can possibly succeed, since any justification of inductive inferences must involve a rule of succession, and the derivation of any such rule will involve equiprobability principles that will automatically provide the basis of a theory of logical probability.

Many of those equiprobability principles—covering, for example, permutations of individuals, families of properties, and equal ranges in the case of quantitative properties—are relatively uncontroversial, but they are not sufficient to yield a sufficiently strong theory of logical probability, and need to be supplemented by either a structure-description equiprobability principle or a state-description equiprobability principle. Both John Maynard Keynes (1921, 43ff.) and C. D. Broad (1927, 6; 1968, 91), however, advanced a very strong objection to a structure-description equiprobability principle, while, contrary to arguments advanced by C. D. Broad (1927, 6–7; 1968, 91–2) and Rudolf Carnap (1962,

564–5), a rule of succession can be derived from a state-description equiprobability principle whose consequences agree strongly with one's intuitive judgments.

Turning to the second issue, Carnap (1962, 570–4) concluded, in effect, that no cosmic generalization could ever be well confirmed. His conclusion was based upon a structure-description-based rule of succession, but the situation is even worse if one uses a state-description-based rule of succession. The view that laws of nature are cosmic regularities, perhaps satisfying some further constraint, therefore leads to the conclusion that one can never be justified in believing in the existence of laws of nature.

One can show, however, that if governing laws are logically possible, then one can be justified in believing in the existence of laws of nature—a result that is due to the fact that governing laws, understood as second-order relations between first-order properties, are *atomic* states of affairs. The problem, of course, is that if Hume was right in holding logical connections between distinct existents *of any kind*—as opposed to just spatially or temporally distinct existents—then governing laws of nature are logically impossible, and nomological skepticism is inescapable.

11. Summing Up

I have set out and defended a non-reductive, theoretical-term analysis of the concept of the relation of causation. That defense involved, first of all, an appeal to several very desirable properties of the analysis itself, including that it is by far the simplest of all analyses, that it immediately entails the correct relationships between causation and probabilities, and that it provides the basis for the only satisfactory account of the relation of temporal priority. Then, secondly, I considered the alternative accounts that other philosophers have defended— Humean reductionist approaches, non-Humean reductionist approaches, and the view that the concept of the relation of causation is analytically basic—and I argued that all of those approaches are open to decisive objections, none of which pose any problem at all for the account I have offered.

References

Bennett, Jonathan, 1974. 'Counterfactuals and Possible Worlds.' *Canadian Journal of Philosophy* 4: 381–402.

Bird, Alexander, 2007. *Nature's Metaphysics*. Oxford: Clarendon Press.

Braithwaite, R. B., 1953. *Scientific Explanation*. Cambridge: Cambridge University Press.

Broad, C. D., 1927. 'The Principles of Problematic Induction.' *Proceedings of the Aristotelian Society* 28: 1–46. Reprinted in C. D. Broad, *Induction, Probability, and Causation: Selected Papers*, 86–126. Dordrecht: D. Reidel, 1968.

Carnap, R., 1962. *Logical Foundations of Probability*, 2nd ed. Chicago: University of Chicago Press.

Demarest, Heather, 2017. 'Powerful Properties, Powerless Laws.' In Jonathan D. Jacobs (ed.), *Causal Powers*, 38–53. Oxford: Oxford University Press.

Fine, Kit, 1975. 'Critical Notice—Counterfactuals.' *Mind* 84: 451–8.

Glynn, Luke, 2011. 'A Probabilistic Analysis of Causation.' *British Journal for the Philosophy of Science* 62: 343–92.

Grünbaum, Adolf, 1973. *Philosophical Problems of Space and Time*, 2nd ed. Dordrecht: D. Reidel.

Hacking, Ian, 2001. *An Introduction to Probability and Inductive Logic*. Cambridge: Cambridge University Press.

Harré, Rom, and Madden, Edward H., 1975. *Causal Powers*. Oxford: Blackwell.

Hume, David, 1739–40. *A Treatise of Human Nature*. London.

Keynes, John Maynard, 1921. *A Treatise of Probability*. London and New York: Macmillan.

Lewis, David, 1970. 'How to Define Theoretical Terms.' *Journal of Philosophy* 67: 427–46.

Lewis, David, 1973a. *Counterfactuals*. Cambridge, MA: Harvard University Press.

Lewis, David, 1973b. 'Causation.' *Journal of Philosophy*, 70: 556–67. Reprinted, with postscripts, in D. Lewis, *Philosophical Papers*, Vol. 2, 159–72. Oxford: Oxford University Press, 1986.

Lewis, David, 1979. 'Counterfactual Dependence and Time's Arrow.' *Noûs* 13 (4): 455–76. Reprinted, with postscripts, in D. Lewis, *Philosophical Papers*, Vol. 2, 32–66. Oxford: Oxford University Press, 1986.

Lewis, David, 1986a. *Philosophical Papers*, Vol. 2. Oxford: Oxford University Press.

Lewis, David, 1986b. 'Postscripts to "Causation".' In *Philosophical Papers*, Vol. 2, 172–213. Oxford: Oxford University Press.

Martin, R. M., 1966. 'On Theoretical Constructs and Ramsey Constants.' *Philosophy of Science* 33 (1/2): 1–13.

Molnar, George, 2003. *Powers: A Study in Metaphysics*. Edited by Stephen Mumford. Oxford: Oxford University Press.

Mumford, Stephen, 2009. 'Causal Powers and Capacities.' In Helen Beebee, Christopher Hitchcock, and Peter Menzies (eds.), *The Oxford Handbook of Causation*, 275–8. Oxford: Oxford University Press.

Oaklander, L. Nathan, 2004. *The Ontology of Time*. Amherst, NY: Prometheus Books.

Paul, L. A., 2009. 'Counterfactual Theories.' In Helen Beebee, Christopher Hitchcock, and Peter Menzies (eds.), *The Oxford Handbook of Causation*, 158–84. Oxford: Oxford University Press.

Paul, L. A., and Ned Hall, 2013. *Causation: A User's Guide*. Oxford: Oxford University Press.

Popper, Karl, 1956. 'The Arrow of Time.' *Nature* 177: 538.

Psillos, Stathis, 2009. 'Regularity Theories.' In Helen Beebee, Christopher Hitchcock, and Peter Menzies (eds.), *The Oxford Handbook of Causation*, 131–57. Oxford: Oxford University Press.

Ramsey, F. P., 1929. 'Theories.' First published in R. B. Braithwaite (ed.), *The Foundations of Mathematics*, 212–36. London: Routledge and Kegan Paul, 1931.

Reichenbach, Hans, 1956. *The Direction of Time*. Berkeley and Los Angeles: University of California Press.

Salmon, Wesley C., 1978. 'Why Ask "Why?"?' *Proceedings and Addresses of the American Philosophical Association* 51 (6): 683–705.

Salmon, Wesley C., 1980. 'Probabilistic Causality.' *Pacific Philosophical Quarterly* 61: 50–74.

Stalnaker, Robert C., 1968. 'A Theory of Conditionals.' In Nicholas Rescher (ed.), *Studies in Logical Theory*, 98–122. Oxford: Blackwell. Reprinted in Ernest Sosa (ed.), *Causation and Conditionals*, 165–79. Oxford: Oxford University Press, 1975.

Tooley, Michael, 2003. 'The Stalnaker-Lewis Approach to Counterfactuals.' *Journal of Philosophy* 100 (7): 371–7.

3

Robust Causal Realism

Evan Fales

1. Introduction

Causation is everywhere, and unavoidable. It is present in determining the motions of bodies. It is present in the workings of the most hidden operations of nature. It is present in the order and arrangement of many sense experiences. Arguably, at least if substance dualism is true, it occurs between mental acts and bodily happenings. Of course, among many theists, it is held to provide the rails upon which contact between the finite and the divine rides. It may fully determine a course of events or only constrain with probabilities.

It is by no means obvious whether these connections between events are all of the same sort, or even whether all of them exist. Here I begin with sense experience, but focus on causal relations in the physical world, setting aside other domains in which such relations might hold. The central claim of my defense of Causal Realism is a rejection of Hume's claim that no such relations are apprehended within sense experience. Hume, I maintain, was wrong about that; but he was right in drawing the inference that, if "necessary connection" is alien to direct experience, then something like Cartesian skepticism about sense experience is inescapable.

My aim here is to argue for a position that can be described as a version of primitivism. It holds that it is a mistake to analyze causation in terms of related predications, such as dispositions, difference-making, subjunctive truths, and so on. Although the view is a form of primitivism, it allows that the attribution of causation in ordinary discourse may cover a range of relations—that "causation" may denote a genus that has species. A central kind of causation is, of course, the kind that we ordinarily attribute to processes we observe in our macroscopic environment. We tend to think of such processes as deterministic, even if the mechanisms lie beyond our understanding. But there might also be indeterministic causal relations—indeed, all physical processes might ultimately be indeterministic. Or they might ultimately all be deterministic. The present view makes no commitments on that score.[1]

---

[1] Besides the deterministic/indeterministic distinction, other arguably fundamental distinctions may be in the offing. One might hold, for example, that the mental processes involved in rational thought are connected causally—but via a causal relation ontologically quite distinct from "physical" causation. Even more radically, one might think, if one thought that there was an omnipotent,

Evan Fales, *Robust Causal Realism* In: *Alternative Approaches to Causation: Beyond Difference-making and Mechanism.*
Edited by: Yafeng Shan, Oxford University Press. © Oxford University Press 2024.
DOI: 10.1093/oso/9780192863485.003.0003

But here, for simplicity of exposition, I shall mean by "causation" just relations of the first two types. Let me call the position at hand *Causal Realism* (CR). It holds that the furniture of the cosmos includes some relation(s), multiply instantiable, that constitute(s) causation. I take the relata of such (two-place) relations to be events. Such relations are instantiated—"really present"—in any particular causal process. I shall presently say more about their ontology.

My argument divides into two main parts. The first is epistemological. It aims to show that we have direct acquaintance in experience with some instances of such causal relations; thus we know that they exist. It then argues that this knowledge bears importantly on the prospects for justifying inductive inferences. How, without begging skeptical questions, can CR account for knowledge of physical properties and laws?

The second question is ontological. I hold that such relations, like other universals, do exist and can exist even uninstantiated, and that causal relations are grounded in a two-place, second-order universal of this sort. That is, causation is grounded in a two-place relation between the first-order properties that suffice to describe our cosmos at the level of fundamental physics.[2] These relations connect those physical properties into a kind of "web" (as it were) of nomological relations. (I shall call it *Charlotte's Web*—though some may prefer to substitute the name of a deity.) But this raises a question about the relation between what I shall call the "whatness" of a universal and the causal relations that exist between it and certain other universals. Is how these causal connections structure (as it were) Charlotte's Web only contingent, or is that somehow determined by the "whatness" of the universals they relate?[3]

The answers to these questions are unclear. So what follows is tentative and my suggestions as to the lines along which answers might be found are provisional.

My plan here is first to recapitulate an argument from the nature of sense experiences to defend the claim that we have direct experience of some instances of causal connection, which give content to our "idea of necessary connection," as Hume misleadingly put it.[4] I make some progress on tackling two epistemological

---

immaterial person who was not a spatio-temporal being but who exerts influence upon worldly events, that some quite different kind of causation must be implicated to connect that being's volitions to events occurring in space and time.

[2] More about the grounding relation presently.

[3] A word about "whatness." The question what kind of thing some particular X is must in general be answered by a list of its properties p_n. But if we raise in general the question, what kind of thing property p_i is, we invite a regress if our only answer can be in terms of a list of higher-order properties for which the same question must be answered. At some point, being of a kind must be ontologically intrinsic.

[4] A more detailed presentation can be found in Fales (1996, ch. 1). There also I defend the essential further claim that causal processes extend beyond phenomenology into a mind-independent world. Indeed, without knowledge of causal relations already in hand, I believe a successful defeat of the skeptic with respect to the senses is out of reach—and with it the primary/secondary quality distinction itself. Direct realism cuts the Gordian knot, but at the cost, in my view, of begging the question against the Cartesian skeptic.

problems: from whence comes our knowledge of causal connectedness; and how far does that knowledge go in enabling one to answer inductive skepticism? That discussion will then make pressing a consideration of the ontology of causation.

2. How Do We Come By Our Grasp of Causal Relations?

The foundation for both the epistemological and the ontological theses concerning causation consists in the argument that some instances of that relation are sensible elements of direct experience—that is, they are constituents of certain of our sensory experiences with which we are *directly acquainted*. In saying this, I am taking on-board a view (with which Hume himself would be sympathetic) that the apprehended contents of sense experience serve, and properly serve, as the ground for our knowledge *that* those contents are presented to us. That is itself controversial, but not defended here.

2.1 The Phenomenology of Causal Experiences

Hume's inquiry into the nature of our conception of cause and effect began, naturally enough, with a careful inspection of the elements of his sensory experiences. And wisely, he chose as his paradigm case a paradigmatically causal process that was as simple and open to inspection as any, devoid of "bells and whistles." In the age of Newton, it was not surprising that his thought-experiment involved envisioning a simple collision between a moving billiard ball and an initially stationary one. That was, however, a fatefully poor choice. We have a number of sensory faculties by means of which to gather news about our bodies and our world. Sight—the sensory faculty upon which Hume's example depended—is one of them. But there are many phenomena it cannot discover to us. It cannot directly reveal the presence of sounds, flavors, odors, temperatures, air currents, pressure, or any number of other significant features that populate experience.[5] What guarantee did Hume have that eyesight should be able to discover the necessary-connection element of causal processes by way of vision? He had none.[6]

---

[5] Of course it can sometimes visually detect these things *indirectly*, as when we infer the temperature from the reading of a thermometer, or the presence of wind from the movements of tree leaves and branches. It is a contingent fact that, for example, color sensations can in most individuals be triggered only by stimulation of the optic nerves, that no color phenomenology results from irradiation by x-rays or infrared radiation, and that the optic nerve is not sensitive to odors. Hume would likewise agree that it is a contingent fact—if it is (as I deny) a fact—that there are objective causal relations, to which our senses are unfortunately unresponsive.

[6] He was not unaware of this, and to catch the omission, he considers (all too briefly) the lacuna in the *Inquiry*, sec. VII, part 1. Hume's counterargument to the felt-force case is stated in ft. 7: 78–9 in the Hendel edition (Hume 1955). This we will consider presently.

It is clear that our olfactory and gustatory senses fare no better than sight in discerning such a connection. But the prospects improve when we consider tactile experiences, aided and abetted by kinesthetic sensations, including those provided by our sense of balance. Here we have experiences of event sequences, recognized as causal, that afford plausible candidates as sources of our grasp of necessary connection. Let us therefore make a close examination of such experiences—most prominently, experiences of pushes and pulls, followed (typically) by experiences of bodily motion.[7]

One way to pump intuitions regarding what elements of a causal sequence are present to tactile and kinesthetic sensory experience is to perform thought experiments. As such thought experiments lie at some remove from what we can actually hope to pursue experimentally, it might be objected that the intuitions they invite are not a reliable guide to what such experiences reveal. But they are no further removed from our actual experience than, say, experiments that appeal to Mary's Room or Swampman. So let's imagine a cousin to Swampman; I'll call him Floatingman. Athena sprang full-formed from the forehead of Zeus; Floatingman forms spontaneously, full-grown, in an apparently empty atmosphere (perhaps from quantum fluctuations). He has never experienced any forces acting upon him: no gravity, no pushes or pulls. No sooner has he overcome the shock of existence than he sees a large object moving toward him from the left at a fair clip. He has no idea what will happen, but he is naturally interested. His anticipation is heightened.

Let's now retreat to the inner sanctum of Floatingman's subjective experience. He has a visual experience as of a large object approaching and colliding with him.[8] He feels a considerable impact to his left side. Alert Floatingman is very quick to make inferences. Our thought experiment asks, what can he rationally infer from this first experience with a forceful push against his left side? Since he has had no prior encounter with forces, he can't appeal to any sort of induction to form an expectation. Would he be able to guess the onset of the sensation of bodily motion? If so, would his expectation as to its direction be just *random*, or would his instantaneous expectation be of the sensation of moving to his right?

---

[7] A number of philosophers (e.g. Armstrong 1997: 216–17) have taken as a paradigm of experiences of causal connection occurrences of acts of willing and consequent bodily movements. But Hume also considers that case and offers a fairly trenchant objection. He takes it that a cause and its proximate effect must occur cheek-by-jowl (i.e. in spatio-temporal contact), and then challenges one to find, for example, the alleged necessary connections between the act of willing a finger to wiggle and the actual wiggling of the finger. For there is no *immediate* connection, necessary or otherwise, between the two observed events. (See Hume, *loc. cit.*) Hume, lastly, considers felt *nisus* (by which I take him to mean felt pushes and pulls), and argues that no necessary connection between them and motion is to be found in those sensations. This I shall proceed to challenge.

[8] Because of his naivete, Floatingman has no way of anticipating what will ensue from mere spatial contact with the rock. Perhaps it will pass through him unfelt; perhaps it will simply vanish or veer away. But he immediately has an additional datum: the thump upon his left side.

Now, with such intuitions, it's everyone for himself; but for my part, there is a strong intuition that Floatingman will expect—and rationally so—the sensations betokening a sudden motion to the right. I think that is the correct intuition. Dedicated Humeans may insist that Floatingman has no rational basis for forming any expectations at all. For them, we may vary the example just a bit. Suppose that the same collision has happened just once before—and Floatingman experienced a motion to the right. What can he expect the second time? A good Humean should answer, I think, that Floatingman should give barely any credence to a prediction of similar motion to the right, as opposed to any one of the infinitely many alternative possibilities. After all, *something* was experienced during the moments following the first collision, but why suppose it had any "advantage"? This is, after all, the first pair of antecedent experiences Floatingman has had opportunity to compare. But surely, Floatingman would be ill-advised not to expect the same outcome.

Such Floatingman intuitions should, if I am right, at the very least prompt us to engage in a close examination of the phenomenology of thumps and tugs. And here, the first feature that should draw our attention is the *vector* character of these experiences. They have a magnitude—there are severe ones and gentle ones—and they have a direction. They point in one direction or another; and this "pointing" isn't something like a visual arrow or, say, the direction from which a sound emanates. While it is not itself a motion, clearly, it is something that does not merely precede or accompany motion, but is felt as direction-giving. It is imperative to avoid two pitfalls here. First, a push or pull—let's just call them collectively forces—do not necessarily entrain motion as a consequent or consequence. Sometimes they occur in the absence of motion: a force may be cancelled by the application of an equal and opposite force. But what is this "cancellation"? It can't be the same thing as the erasing of both forces; there is a difference between a force-free object and one being compressed or under tension from opposed forces. That suffices to show that the connection between an applied force and motion is not a necessary one; but does not even begin to establish that it is merely contingent.

On the contrary: it beckons us to infer that simultaneous application of multiple forces will result (and not just by accident) in some motion that is responsive to their combination. But how do they "combine"? Fortunately, an answer is ready to hand: multiple forces combine to produce a "resultant" force that is the vector sum of the component forces. Nor is this a result that can be inferred only from a wide range of experiments; it can be apprehended, at least in rough form, in the experience of single trials. This requires some attention, but I think that, from any experience of equally strong and opposing forces, one can infer no motion. And from, say, the experience of two forces applied at angles to one another, one can form a rational expectation of a direction of motion intermediate between those of the two forces. The mind, that is, can intuit from experience of single forces something like a necessary rule of composition of forces, albeit at best only qualitatively. And indeed, that intuition is reflected in our understanding that causes combine in the ways modeled in J. L. Mackie's definition of INUS conditions.[9]

I have proposed that the sensory capacities that afford us detection of (instances) of causal relations are our sense of touch and our kinesthetic senses, and that these are responsible for our experience of felt pushings and pullings. A classical foundationalist would say more precisely (as I will) that our experience of causation discerns things that happen in the arena of bodily phenomenology. Thus, they appear in connection with, for example, sensations of bodily motion accompanying the sensations of having, for example, an arm or a torso or a head. For me, then, there lies at the heart of an epistemology of perception the classical problem of skepticism regarding the senses. I want to emphasize that there is no claim here that touch and kinesthesia are sensitive to *all* instances of causal relations between events given in experience. Most must be inferred. But the essential (and I think irreplaceable) role of experience of causal relations is to provide us with what we need to form a robust *conception* of causal relations and processes, and of the kind of explanatory understanding they confer. They offer an answer to Hume.

Most contemporary philosophers are disinclined to regard the thought experiments I have described as dispositive. They may regard them too far removed from actual experience to teach us much. They may say that, whatever experiences Floatingman might intelligibly have can be atomized into discrete, contingently arranged elements between which no necessary connections can be discerned. They deny, in effect, that the sensation of being pushed or pulled bears any necessary connection to that of motion. Of course, my argument relies upon a recognition that the sensation of being pushed involves an irreducible connectedness to motion, and that it is this recognition that provides the intuitive force of the examples.

But there is, I suggest, another reason why perception of causal relations proves elusive. It is that the phenomenology of apprehension of relations *generally* presents puzzles. Some relations are internal relations; others external. An internal relation is a relation that holds in virtue of intrinsic (i.e. monadic) properties of the relata: thus, crimson is darker than pink in virtue of what it is to be crimson and what it is to be pink. Nothing over and above the nature of the two monadic properties need be recognized in order to apprehend that the *darker-than* relation holds; no further relational property—at least no further first-order

---

[9] While Floatingman's experiences satisfy Ducasse's (1926: 59–63) definition of a cause (since they involve a "single change in a situation S being followed by another single change"), we are in no way using our thought experiment to suggest that such a sequence in itself constitutes it as causal, let alone captures the sense of causal necessity.

relation—need be posited. External relations are however *not* "free lunches" once monadic properties of the relata are specified. They do seem to demand a distinct relational property. Relations between spatial positions appear to be of this sort. When we say that billiard ball A lies between balls B and C, we are positing an exemplification of a three-place *betweenness* relation, one that is surely discernible by way of sense experience. But where, exactly, is that relation situated? Is it visually apprehended to attach to the three balls themselves? Or to their ordered triple? Or to the intervening spaces? I suggest that the puzzlement that we feel here over the phenomenology of the *betweenness* relation extends to perceptible external relations generally; and that, if causal relations are external—as I think they are—this fact may in part explain the apparent mysteriousness of their givenness in sensory experience. This, I hope, will suffice to motivate an effort to see whether the claim of givenness can be developed in fruitful ways.

2.2 From Evidence to Hypothesis

The recognition that causal relations, as we encounter them within experience, can obtain between events outside of experience[10] has major epistemological consequences. Among the most eminent are the contributions CR can make to our understanding of the justification of ampliative reasoning—reasoning from known effects to causes, and from causes to effects. These undergird, *inter alia*, foundationalist responses to the problem of perception and the problem of justifying claims regarding laws of nature. Abduction is in my view the more fundamental of these. CR can also shed light on the sources of fallibility in the acquisition of knowledge about the world, and uphold a principled distinction between the "directly observable" and the "theoretical" or hypothetical, and thus provide foundationalism itself with an essential foundation.

The CR strategy, thus, begins with the phenomenally given, specifies a principled distinction between foundational knowledge—knowledge respecting the phenomenally given—and knowledge that requires inferences about what causes what. Having identified repeating features of experience, it provides tools for identifying mind-independent properties and, consequently, substances.[11] Nevertheless, CR does not thereby defeat skepticism. For one thing, as we have seen, the identification of physical properties is subject to uncertainties. Those uncertainties reflect the problem of underdetermination of "theory" by data. That is a problem that remains to be addressed.

[10] See Fales (1996: 39–46).
[11] Using patterns of experience that suggest common causes; see Fales (1990: 193–219).

3. The Ontology of Causal Relations: Preliminaries

3.1 Form and Ground

A substance having a property is an event/state involving that substance and its first-order properties, or n substances instantiating a first order n-ary relation. Yet the root of the causal relation goes deeper than relations between events. An indication of this is given in the generality of laws of nature. The laws—at least the fundamental ones—do not mention particulars; they state only relationships between properties. That suggests, correctly, that causation is grounded in a relation—now a second-order relation—between properties. It is this second-order relation that provides the ontological foundation for causality. When an event A causes another event B, we have an instance of (second-order) causation. Certain (first-order) properties of A and B (call them a and b) instantiate the second-order causal relation (call it C); the derivative first-order causal relation between A and B is in turn an instance of aC b. It is, then, an instance of an instance of C.

3.2 Laws and Universals

This structure has an immediate consequence for those laws of nature that express causal relationships.[12] It is that those laws have their ontological ground in the second-order causal relationships between physical universals. The properties they relate are not the children of convention, but discoverable in nature. CR should accept property realism, at least for the properties required for an adequate description of nature.[13] Each of them will have a distinct "nature" or character that is not reducible without remainder to the possession of second-order properties. For if a property's character were entirely a matter of its possessing higher-order properties, the same question would arise concerning the character of those higher-order properties, and the resulting regress would not be benign.

But how strong is the connection between a first-order property p and its C-relations to other first-order properties? Does p have such relations essentially or accidentally? Some modal intuitions support the former. Is it possible for the negative electric charge of an object to attract, rather than repel, other objects

---

[12]   There are laws that directly express causal relationships, such as the law that a 2:1 ratio by gram molecular weight of hydrogen and oxygen at atmospheric pressure and 100 °C will ignite explosively if subject to an electric spark, and ones that summarize a myriad of causal processes, such as the Boyle-Charles Gas Law ($PV = nRT$). There are also laws of nature that are non-causal: for example, that all electrons carry a charge of $\sim 1.06 \times 10^{-19}$ coulombs. We are not concerned here with the latter.

[13]   These will include mathematical properties, though they are not themselves the relata of causal relations.

in virtue of their also being negatively charged? Would it be possible for a spark-ignited mixture of hydrogen and oxygen tranquilly to transform into a small, living frog? If these were so, could we still speak, with propriety, of two *negative* charges, or of genuine hydrogen and oxygen? Just as one might plausibly argue that photons are essentially elementary particles with zero rest mass and no electrical charge,[14] it seems right to say that it is part of what it is to be negatively charged that two things charged in this way exert a repulsive force upon one another, and part of what it is for a particle to have zero rest mass that it travels at the speed of light. I shall say that the C relations to other properties had by property *p* are, by analogy, essential to it. They hold necessarily, and hence, underwrite the necessity of fundamental physical laws. This, in turn, bears on other issues.[15]

4. Knowledge of How the World Works

The first question embraces the problem of justifying the kinds of reasoning that lead to inferences about the laws of nature, reasoning that is by all accounts ampliative. A natural strategy for the Causal Realist is to favor abductive reasoning over, say, enumerative induction as offering the better view of the most funda-mental route to non-deductive knowledge or, at the very least, as a required supplement.[16] With an adequately explicated conception of causality in hand, the realist can provide a substantive answer to the question why causal explan-ations are superior to "brute repetition" conceptions of laws of nature, and so better explain observed regularities. We can see why, once a causal relation is identified between universals, there follows a law of nature. But how is such identification to be achieved?

There seems to be a tight circle between identifying a universal and identifying the causal relations that tie it to other universals. To identify the former (except in the case of phenomenal properties, whose natures are immediately apprehended), we must discover the unique "fingerprint" of causal relations it bears to other universals; but to determine that fingerprint, we must nail down the identities of these related universals. Obviously a regress threatens, and equally obviously, our only option appears to be to begin with identifying phenomenally given proper-ties, for which we need no such fingerprints, and building outward from there. But our store of distinctly identifiable phenomenal properties is most likely finite; and even if it were not, the number of phenomenal property-instances that we can effectively reason from certainly is. Yet there are very likely continuum-many determinate physical universals, as most of our understanding of relations

---

[14] As I would; Fales (1982).   [15] But see the section on necessary causal connection below.
[16] See Weintraub (2013).

between physical properties invokes continuous functions.[17] How can there be any hope of pinning down the strands of a web of causal relations that involves continuum-many threads, given a finite set of data upon which to rely? Putting it more succinctly, can scientific investigation, in principle, hope fully to discover the underlying laws that make the universe "tick"—a Theory of Everything (TOE)?

Duhem's theorem poses a general formulation of the problem. Eliminating all competitors to a TOE is correlated to solving the puzzle of pinning down the strands and nodes of Charlotte's Web. Underdetermination of theory by data also threatens a kind of "halting problem": can a finite data pool D ever be assembled that will decisively limit the number of viable theories? Or can the construction of alternative theories be indefinitely prolonged? More precisely, regarding a TOE, is there any prospect of assembling a D that will so strongly favor a given theory as to assign it a level of confirmation higher than that of the disjunction of its possible competitors?

It is not obvious that Duhem's theorem confronts us with an infinite quest. Nor is it obvious, I think, that pruning away branches requires appeal to problematic criteria of simplicity, dubiously justified prior probabilities, or the like. For the theorem does not license unrestricted tinkering with auxiliary hypotheses when a theory faces failures of prediction. The constraints derive precisely from the web-like structure that the theorem itself trades on. First, adjustments must face the tribunal of independent testability, and survive the tests. Second, they must not lead to incoherence in the larger body of theory. The problem, then, comes to this: is there a unique TOE that is internally coherent, survives continued independent testing, and sufficiently out-strips competitors in predictiveness?

I need to say more about coherence. I have in mind something more than mere logical consistency. What is at stake is a kind of explanatory unity. The idea has received considerable attention;[18] here it will be sufficient to illustrate what is meant by discussion of a particular kind of case.

Consider the curve-fitting problem. There is, to take a simple example, a functional relation between the temperature of a volume of gas and the pressure it exerts upon a rigid container, given that other variable characteristics are held constant. We have a finite set of measurements of this correlation, which suggest that it is roughly described[19] (up to experimental error) by a continuous function: $PV = KT$, where K is a constant. But, of course, those data points are fit by also a

<hr/>

[17] I am assuming here that space and time are continua. This is unsettled. Whether there are laws that are continuous functions of physical variables would require re-consideration in the contrary case. The frequency of electromagnetic radiation apparently varies continuously, but if space-time were quantized, that might entail that possible frequency values do not form a continuum.

[18] See, for example, Philip Kitcher (1989). I am not depending here, however, on Kitcher's development of the idea.

[19] That is, ignoring van der Waals forces, which become significant near phase transitions.

non-denumerable infinity of other curves, which pass through those points in P/V/T space no matter how many there are. So those functional relations are not falsified.

One response to this difficulty is to appeal to the intuition that the other curves are not predictive; at values of T incompatible with the inverse function, they give values of P that are incorrect when tested. Other curves steadily fall by the wayside when challenged. But this ignores the point that infinitely many alternatives survive any given challenge. And, indeed, Bayes' Theorem does not care if a successful "prediction" is made before experimental challenge or in hindsight.

A more successful point concerns explanation. Consider another functional relation that deviates from the Boyle-Charles function above in a region not pinned down by data. Chances are excellent that it will not be explainable or consilient with broader theory, whereas the Boyle-Charles relation does mesh with broader theory—for example, with conservation laws. More trenchantly, the Boyle-Charles relation followed naturally from the much broader considerations of statistical mechanics, once these were developed; our competing hypothesis almost surely will not. These are, precisely, holistic considerations derived from a very wide range of empirical results and attendant plausibility-considerations. Indeed, a few competitors, such as the van der Waals gas law, which improve upon the Boyle-Charles law, are directly inspired by the consideration that the derivation of the latter from statistical mechanics involves the making of *un*realistic simplifying assumptions which can be replaced with more realistic ones.[20] So ancillary considerations, motivated by independent empirical evidence, can point in directions that allow better fits to data.

One might hope that our data, taken as a whole, in conjunction with the kinds of constraints imposed by scientific reasoning illustrated here, would suffice to offer substantial progress toward a true TOE. Unfortunately, I have no argument to show that this is so. Still, the gas-law story does offer a hint that provides some grounds for optimism. Our discussion traded on the explanatory power of reductive explanations (here, the reduction of thermodynamics to statistical mechanics). Explanations of the behavior of wholes in terms of the behavior of their parts, though not a universal strategy, is ubiquitous in the sciences because of its power. And current science provides a good deal of support for the expectation that at the heart of a TOE will be a characterization of the properties and laws that describe some set of elementary—that is, part-free—bits of matter out of which everything is composed.

It is well understood that reductive theories can be dramatically unifying in their explanatory scope. It is not unreasonable to hope that a satisfactory account

---

[20] The Boyle-Charles law assumes that the gas molecules collectively occupy none of the volume of the gas, and that collisions between them are perfectly elastic, involving no attractive forces. Van der Waals' law, which rejects the simplifications, gives significantly better empirical predictions near the gas phase transition points.

of the elementary bits of matter, the fundamental forces that animate them, and the fundamental nature of space-time would in principle mark the achievement of a TOE. It is more speculative, but also not unrealistic, to think that the number of elementary particles—at least those that can exist at physically achievable energies in our universe—and the number of elementary forces will be relatively small; indeed, enormously simpler than the complexity of the universe built out of them. If, indeed, the number of elementary particles is not more than, say, a few dozen, the number of elementary forces not larger than, say, four, and the dimensions of space-time not very high, then pinning down a TOE to a single viable candidate or just a few viable competitors might not be unrealistic to hope for (given continued scientific investigation). That is to say, the nodes and connections that specify the web of fundamental properties and causal inter-actions might be discoverable.[21] That would suggest that, equipped with a robust realist conception of causation, a realist conception of properties as universals, and an evidentiary base that builds upon direct acquaintance with the phenomenally given properties, it may be within reach, in principle, to arrive at a well-justified TOE, one of the central aims of scientific inquiry.

But perhaps this is entirely too hasty a conclusion. I have described Charlotte's Web as specifying essential (second-order relational) properties (first-order) physical properties. That would entail that, once the physical properties of the universe are fixed, the laws of nature are as well—a reassuring result.

5. The Ontology of Necessary Causal Connection

But the fixing of the laws is not something that requires that there be some sort of necessary connection between the intrinsic nature of a physical property and these causal relations that fix its location in the web.[22] If that is a matter of necessity, what sort of necessity is it? And how, or why, does the nature of a property necessitate the congeries of essential causal relations? This is, indeed, the second

---

[21] I am, to be sure, conveniently setting aside here problems posed by a number of conundrums that may threaten this irenic view of scientific progress, such as the alleged conventionality of the one-way speed of light, the alleged threat that the measurement problem in Quantum Mechanics poses to scientific realism, and the existence of "unphysical" mathematical results that must be either discarded or rather artificially circumnavigated.

[22] This interest in nomological relations being essential to the related universals is motivated in part by the question whether two (or more) distinct universals, U and U*, could have identical nomological profiles—hence, occupy the same position in Charlotte's Web. If so, they would be indistinguishable by us if non-phenomenal (since we only identify them empirically via engaging their instances in causal interactions with our sense organs), though differing in "whatness." The existence of both U and U* (rather than just one of them) would make no difference to physics—so one might not care. But if we can say that the complement of causal relations that characterize a universal comprise its entire essence (whereas the "whatness," not being a property of the universal, is not itself strictly a component of its essence), then essence and identifiability are so tightly connected that joint tenancy of nodes in Charlotte's Web is precluded: same essence, same "whatness."

puzzle that I want to discuss. The "connection" is, evidently, an exemplification "relation"—and hence not, ontologically speaking, on a footing with "ordinary" relations (of any order). If it is necessary, its necessity is, one would presume, a metaphysical necessity: such connections are not specified by analytic truths and could only be knowable a posteriori. But if I have been right about the epistemic status of causal relations themselves—viz., in claiming that in certain cases we have direct acquaintance with exemplifications of them, what are we to say about epistemic access to the "relations" presently at issue? How is it that the nature of a universal dictates second-order causal relational properties?

A first thought might be that we are, perhaps with certain phenomenal properties, directly acquainted with the necessary connections that determine at least some of their causal relations to other universals. Like good Humeans, we might examine experience in search of such instances, in the hopes of obtaining direct acquaintance that provides the sought-after conceptual content. But I fail to discover such constituents in the phenomena of experience. It is not as if nothing in the way of necessary relations is discoverable between the natures of phenomenally given universals. The clearest cases would be internal relations: relations that are integral to the natures with which we are acquainted. In this sense, pink is lighter than crimson. We recognize that simply by recognizing their respective given natures. But, as Armstrong (1978: 10–11) notes, this sort of relation appears to be an ontological "free lunch": it requires no postulation of any entity beyond the natures themselves. Could the necessity of causal relations having the relata they do be like that—implicit in the natures of the related universals? If so, this does not appear to appear in the appearances.

It could however be that we should not infer from this that causal relations are not internal relations between universals. For one could allow that there can be more to a property that is phenomenally given than what is presented by the phenomenal content. Might the phenomenal content reveal only some *part* of the "nature" or "whatness" of a property? To allow this would be to deny that there can be nothing more to a phenomenal property than what can be discovered by direct examination of experience of it. It would be to see direct acquaintance with universals as perhaps only a partial revelation.

One might infer such a conclusion anyway from the fact that universals are not typically given in experience as fully determinate: to say they are is to invite sorites paradoxes. (Thus, two patches of phenomenal pink might appear indistinguishable in color, although in fact different but too closely resembling to be distinguished.)[23] What's worse for CR, dialectically, is that it would threaten realism itself by construing causal relations as an ontological "free lunch."

---

[23] I am not myself persuaded by this sort of case, and hold that what it shows is that many universals given in experience are not fully determinate; see Fales (2010: 73–7).

The path forward is, then, rather to treat causation as grounded in a second-order external relation that is exemplified by first-order physical universals. And this leaves unresolved the original question why these relations hold between universals in precisely the way they do, rather than in some other pattern. What is the ground of their necessity?

Perhaps a bit of insight can be gained by considering the ancient question of what distinguishes the essential properties of a particular from its accidents. By the early modern period, philosophers had become widely skeptical of the claim that there was any such distinction to be had. In what way was an essential property "anchored" to a particular differently from the (pseudo) relation *exemplification* that obtains between a particular and its accidents? And anyway, how are we to defend taxonomic essentialism—for example, in the case of animals and plants— given the existence of hybrids and mutants?[24]

Now in my view, this is a misleading way to frame the problem. For, I should say, there are no true species and no hard taxonomic boundaries that license anything beyond rough pragmatic species-distinctions for living organisms. But that should not come as a surprise, for organisms are enormously complex collocations of molecules, and thus by their very nature vulnerable to sorites difficulties. The picture changes as we travel down the reductive chain to simpler forms of matter: first molecules, then atomic species, and finally elementary constituents of matter. Now taxonomy of molecules and atoms is also complicated by problems regarding the fixing of "true species": consider chirality in the case of molecules and atomic number (isotopes) in the case of the chemical elements. But the situation becomes clearer (because offering fewer degrees of freedom in varying characteristics) as we move from the complex to the elementary. And *if* nature is so organized that the reductive hierarchy terminates in elementary particles or the like, then the difference between essence and accident becomes stark. The essential properties of an elementary particle (or field) are just its *monadic* properties, and the accidents are just its *relational* properties. And these, it now appears from symmetry considerations, are very few: spin (in units of $\frac{1}{2}\hbar$), charge, and rest mass. The fundamental relational properties might also prove to be small: space-time relations and (at the second order), the causal relation C. The fundamental forces appear to be roughly four (possibly just one), as we noted, and the number of fundamental particles—at least those whose rest masses are low enough to allow creation in our universe— may number in the tens.[25]

---

[24] See Locke (1959: 56–84; bk. III, ch. IV, §1–§32) on these problems and others as grounds for skepticism.

[25] I have put the point here in terms of fundamental particles. It is more common at present to treat the four known fields as fundamental. This would complicate the exposition somewhat, but the essential point remains the same. The fundamental fields have certain intrinsic features (including, e.g., spin and "bare" mass) that distinguish them and that are independent of reference frame, as well as

Now all this is not only tentative but is itself a finding of theory, not simple data. But that does not undermine the point here: if the hand-in-hand development of theory and data confirms that the strongest theories we have point toward ultimate explanations that appeal to only a very limited number of elementary building-blocks, then we have reason for optimism that a TOE will follow suit.[26] This, in turn, supports optimism that progressive reductions will put progressively tighter constraints upon admissible theory and ultimately a finite limit upon the number of competing theories that remain viable.

As a simple thought experiment, we might imagine a TOE positing just four genera of elementary particles whose interactions are governed by two forces that push or pull them. The forces depend upon the types of particles interacting, and so upon their monadic properties and their (contingent) spatiotemporal relations. Our question is: what is it about a pair of particles having mass that results in (say) an attractive force between them, and what is it about their both being positively charged that results in a repulsive force? Or are these simply to be taken as brute (though necessary) facts?

An initial suggestion can be quickly dismissed, viz. that perhaps the answer to this question is to be sought by empirical investigation—by some sort of experiment that would detect the necessary connection we seek. It does not seem that this suggestion can bear fruit. The necessary connection is an exemplification "relation" that holds necessarily, given the intrinsic "whatness" of mass. But any experimental detection method of the usual sort requires a measurement that is a causal process. So it would require that the exemplification itself display necessary causal connections to the event that initiates that process. But this invites a regress, and the regress is vicious. For what grounds the causal connection of the target exemplification to the measuring event will include the "whatness" of (second-order) exemplification? We are left both with such exemplifications being relations that have a "whatness" and with the question of how that "whatness," whatever it is, provides the ground for the necessity of the causal relation that initiates the measurement.

But this way of looking at the matter has led into a trap. It is premised on the hope that something about exemplification can explain the difference between necessary and contingent possession of a property, and that lures us into thinking of exemplification as itself a relation that can do explanatory work. I have already suggested that, in the case of particulars and their properties, the difference

relational properties. Its monadic features make a field what it is—for example, a photon (electromagnetic) field rather than a gluon (strong-force) field. As for the elementary-particle view, current taxonomy recognizes 17 species, falling into two families (fermions and bosons).

[26] The ancient Greeks imagined a world composed of four basic elements; after hundreds of years, we now subscribe to a periodic table that accounts for the structure of ordinary matter by roughly 100 elements and 17 elementary particles. The fact remains, however, that explanations of wholes in terms of parts has the intrinsic possibility, and tendency, to explain more in terms of less. If reductive explanation has a terminus, it is a fair bet that it will appeal to a small number of elements.

between essential and contingent properties boils down ultimately to just the difference between monadic properties and relations. Here, we are considering properties themselves and one of *their* relations. Might all simple (i.e. ultimate) second-order properties be essential to the first-order properties that exemplify them?

A first important point is that a Platonist holds that universals are not temporal entities. So it is not coherent to think that they might change in time. This applies to the complement of causal relations a physical universal has; changes in their C relations would not be "Cambridge" changes. That secures the result that the fundamental laws of nature are not subject to change. But could the membership in such a complement have been nevertheless contingent "at the outset"? If it is intelligible that a god could have decided with which physical universals to populate its universe, might that god have settled upon a different complement of causal relations for any given universal—thereby bringing into being a universe with perhaps the same physical properties as ours but different laws of nature? I do not know how to answer this question, in part because I do not know what sense can be given to the creation of a timeless entity.

Inquiry and explanation must stop somewhere, and if it stops here vis-à-vis causal necessity and the laws of nature, I should not be too disappointed.

6. Back to Epistemology: Knowledge of the World

It will be evident that CR has a good deal to suggest about empirical access to the world. Let me return briefly to the relationship between experience and the world that CR offers. If we wish to attempt a systematic reconstruction of empirical understanding of the physical world, I suggest that knowledge of physical properties is the place to begin. It is, obviously, by recognizing the properties of a thing that we can identify it and come to know about it.

Acquiring knowledge of physical properties that inhabit Charlotte's Web is obviously a piecemeal and fallible process. Our knowledge of such properties is in general unavoidably indirect: it is a matter of identifying the powers a property confers upon those particulars that instance it, powers that can be displayed only in particular circumstances and hence by varying observations that "triangulate," via causal chains, the property (instances) in question. What distinguishes a property, from our epistemic position, is its associated C relations to other properties—the powers that possession of it confers. But two properties may confer many of the same powers; ultimately what distinguishes them might be only one power conferred by one but not the other. Absent discovery of that difference, we will not be able to distinguish between these properties.

How can this process of nibbling away at uncertainties even begin, unless we can identify the universals to which a mystery universal is Cly related? The answer, pointed to above, is that one must begin with directly recognizable properties, properties identifiable, not in virtue of their C relations, but as what they intrinsically are—that is to say, phenomenal properties of experience—and then to deploy these to zero in on the C relations of other properties.[27]

7. Some Concluding Reflections

I have been arguing that CR provides a powerful tool in the contest between skepticism and realism. Perhaps it can offer some advance to the problem of justifying ampliative reasoning. Here again, I shall have to be brief. The basic idea, one that has been defended by a number of philosophers, is that ampliative reasoning is at bottom reasoning to the best explanation. Whereas enumerative induction is, in typical examples, reasoning from known causes to effects, abductive reasoning is in its element in seeking causes of known effects.[28] My suggestion is that the latter is more fundamental. The reconstruction of enumerative induction is then possible. Faced with a pattern in our data, we ask whether it is better explained as coincidence or causation—that is, a non-accidental connection that holds between paired events and does so in virtue of similarities between the pairs. The latter hypothesis wins this contest—or rather, does so if a relevant connection can be identified or plausibly inferred in the light of our other theoretical understanding.

In some such cases, a necessary connection is both evident in experience and of the right sort to provide an understanding of why the pattern is not a matter of chance. In other cases, direct experience of such a connection is absent, but conjecture that it is nonetheless objectively present provides an explanation of the pattern not otherwise available.

Hume would argue that we have "no idea" (or at most a negative idea) of such objective necessary connections. I can flesh that out by pointing out that such a postulation has a very different status methodologically than that of our usual theoretical posits. Characteristically, theories posit features of the world that

---

[27] Having knowledge of the causal relation itself in hand, and noting that we can recognize the presence of phenomenally given forces (e.g. gravity) even in some instances in which their source remains outside the phenomenal field, we reason that mind-independent causes must account for them.

[28] See Weintraub (2013: 213–14). Weintraub suggests that abductive reasoning ultimately falls victim to a skeptical objection parallel to the one Hume mounts against enumerative induction: that, analogous to the required assumption of a question-begging Uniformity of Nature postulate in a justification for the former, one needs a "Principle of Universal Explicability" to justify the latter. But the two justifications are not parallel: explanation, arguably, must end somewhere, whereas a demand that regularity must "end somewhere" has no clear sense. Explanation might end in something that is a brute fact—that is, has no explanation—or in something that is in some sense self-explanatory or not requiring explanation. And explanatory stasis might be reached either because a phenomenon has no best explanation or because it has no explanation at all. But that explanation must reach an explanation-bereft terminus does not entail that the unexplained cannot be an explainer.

explain data by providing a causal explanation for those data. Those explanations presuppose that the causal relation is already understood. But the positing of causal relations themselves can't help itself to *that* kind of explanation; it must appeal to some unfleshed-out notion of necessary (or rather non-accidental) connection that generates all the structural characteristics of causal processes (e.g. INUS conditions, pre-emption, finks, overdetermination, and the like). The power of CR is illustrated by the ease with which such difficult cases for other theories (reductions to counterfactuals, powers, possible worlds, and the like) are constructed; they are cases guided by our natural intuition that causal connections are real relations between events that can act jointly in myriad ways.

The quest for laws of nature and theory confirmation rests, then, on a robust process of reasoning to the best explanation. That is, of course, no trivial matter. A realist understanding of causal connection as binding events together in a way that is clearly contrasted to mere brute succession can contribute to that project, however. For it enriches our conception of what count as better explanations and as worse ones. It does that in part by giving us certain ideal models of causal interaction: for example, it gives a certain intuitive preference to models of force transmission via contact, where these are available, and shapes our thinking about action at a distance. It gives us a vivid picture of how causes can act collectively in determinate ways. It provides clear intuitions about the way understanding of forces is essential to understanding motion. And it provides an ontological basis for the observation that laws of nature are not islands in Platonic heaven, as it were, but form a systematic and interlocking template for the doings of our world.

The identification of causal relations with forces whose nature is occasionally an object of direct acquaintance is, then, something I hold can be settled by experience itself; and dialectically, that acquaintance provides us with a fundamental conceptual tool in the defense of both epistemic realism and metaphysical realism respecting Platonic universals.

In saying this, I do not mean to be recommending CR as a panacea against all forms of skepticism, by any means. The road from knowledge of non-accidental single-case causal connections to a robust theory of confirmation remains long and littered with difficulties. But I hope the above observations regarding the constraints on theory change offer some reason to think that an empirically satisfactory TOE may not be in principle out of reach.

References

Armstrong, David M. (1978). *A Theory of Universals*. New York: Cambridge University Press.

Armstrong, David M. (1997). *A World of States of Affairs*. Cambridge: Cambridge University Press.

Ducasse, C. J. (1926). "On the Nature and the Observability of the Causal Relation." *Journal of Philosophy* 23 (3): 57–68.

Fales, Evan (1982). "Natural Kinds and Freaks of Nature." *Philosophy of Science* 49 (1): 67–90.

Fales, Evan (1990). *Causation and Universals.* New York: Routledge.

Fales, Evan (1996). *A Defense of the Given.* Lanham, MD: Rowman and Littlefield.

Hume, David (1955). *An Inquiry Concerning Human Understanding.* Edited by Charles W. Hendel. Indianapolis, IN: Bobbs-Merril. Originally published in 1748.

Kitcher, Philip (1989). "Explanatory Unification and the Causal Structure of the World," in Philip Kitcher and Wesley Salmon (eds.), *Scientific Exploration* (Minnesota Studies in the Philosophy of Science, Vol. 13), 410–505. Minneapolis, MN: University of Minnesota Press.

Locke, John (1959). *An Essay Concerning Human Understanding*, Vol. 2. Edited by Alexander Campbell Fraser. New York: Dover Publications.

Tooley, Michael (1997). *Time, Tense, and Causation.* New York: Oxford University Press.

Weintraub, Ruth (2013). "Induction and Inference to the Best Explanation." *Philosophical Studies* 166: 203–16.

4

The Metaphysics of Causation

An Empiricist Critique

John D. Norton

1. Overview

This chapter presents an empirically based, skeptical critique of the metaphysics of causation. The mere mention of empiricism is apt to lead to a misreading of the approach. It is not a Humean critique. Hume's celebrated critique of causation was a mixture of inductive skepticism and causal skepticism. He doubted that we could know of causal connections because he doubted the inductive inferences that offered that knowledge. I do not share Hume's inductive skepticism. It was based on an impoverished and unsustainable conception of inductive inference. In its place, I am an inductive optimist and have mounted a far-reaching defense of inductive inference that includes what I believe is a quite serviceable dissolution of Hume's skeptical attack on inductive inference (see Norton, 2014; forthcoming, ch. 6).

My inductive optimism extends to the connectedness of things in science. I have no dispute with the idea that we have learned inductively of many ways that things connect in the world. We have good reason to accept that electric currents are set in motion by electric fields; that chemical reactions are driven forward by gradients of free energy; that natural selection leads to a diversity of species adapted to their ecological niches; and so on and on over the full range of our sciences.

Call these last connections "causal" if you like. In my view, attaching the label "causal" has only pragmatic uses such as will be developed below. It may be little different from the practice of florists when they declare some plants flowers and others weeds. There is no deeper basis for the division in botany. However, it is a definite fiscal benefit to the florists to offer their customers what the customers will identify as desirable flowers. Or it may be akin to the way chemists divide substances into organic or inorganic. There is no prior necessity that the world must provide us substances of either type. However, once the division is found empirically, it is useful since chemists treat each type of substance by different methods.

John D. Norton, *The Metaphysics of Causation: An Empiricist Critique* In: *Alternative Approaches to Causation: Beyond Difference-making and Mechanism*. Edited by: Yafeng Shan, Oxford University Press.
© Oxford University Press 2024. DOI: 10.1093/oso/9780192863485.003.0004

The danger with causal labeling is that it invites confusion. Calling a process[1] "causal" is too often understood as identifying it as a manifestation of a factual, causal order in the world that has been identified by prior metaphysical analysis. The present study of the *metaphysics* of causation, I shall argue here, has nothing useful to add factually to the catalogue of the connections discovered by science. Rather the metaphysical investigations of causation are confused in their foundation and purpose. They proceed under the illusion that diligent reflections on the meaning of terms and drawn-out investigations of how we ordinary folk use them can somehow provide a deeper illumination of the causal nature of the world. These efforts are, in my view, a failed attempt to pursue a priori science.

The principal means for developing the argument of this chapter is in a dilemma, posed in Section 2: either conforming a science to cause and effect places a restriction on the factual content of a science; or it does not. The first horn is explored in Section 3. It shows that the history of efforts in the metaphysics of causation to place factual restrictions on processes in the world is one of sustained failures. Earlier, failed maxims of causation were replaced in the nineteenth century by the single idea that causation is just determinism. When that dogma failed with the discovery of the indeterminism of quantum processes, it was replaced by a new dogma: that causation is probabilistic. This new dogma is now widely accepted, but it is ill-fated. It rests on the false assumption that all indefinitenesses in the world are probabilistic. In sum, this first horn is rejected; and the second horn accepted.

The second horn of the dilemma is developed in Sections 4 and 5. If causal analysis cannot provide novel factual restrictions on processes, then it is reduced to an exercise in labeling. We are free to choose how we might like to assign these labels. Done poorly, the assignment is unsystematic. If it attempts to mimic ordinary causal talk, the result will be inconsistent, since the ordinary use of causal language is inconsistent.[2] Done well, the exercise can have definite utility. Once we conceive a process in causal terms, we connect it with others we find analogous. The result is commonly an enhanced comprehension of the processes. The choice of causal labeling can be especially helpful if it tracks useful factual distinctions in processes. In Section 6, my version of the interventionist account of causation is identified as such a case. Its causal labels distinguish those processes that admit interventions. We thereby learn of processes that allow us to manipulate the world. This is useful knowledge. In Section 7, I show how this reduced conception of interventionist causation protects it from standard objections: that

---

[1] Here and below, the term "process" is used informally merely to designate a part of the world of interest as it undergoes changes.

[2] Causation by omission occurs when the absence of a factor means that some process is not blocked. For some, the absence causes the process. For others, it is merely blamed.

the account does not apply where interventions are impossible; and that it is circular since it takes intervention itself to be a causal notion.

Modern writing in the metaphysics of causation has become noticeably reluctant to propose causal maxims with factual content. An exception is recent work that seeks to impugn the notion of downward causation. This type of causation is defined as arising between different levels of description of the world. Common examples are proposals of causal processes from the mental to the physical in theories of mind; and from the thermodynamic to the statistical mechanical levels in thermal physics. Section 8 introduces the notion of downward causation and agrees with Woodward that downward causation is supported by an interventionist account of causation. Section 9 reviews the causal maxims that have been proposed as refuting the notion of downward causation. The maxims include the ideas of causal exclusion and closure; the prohibition of whole to part causation and synchronic causation; and the requirement that causal relations must be asymmetric and proceed forwards in time. In a continuation of the project of examining the first horn of the dilemma above, the section shows how each maxim fails and thus fails to impugn an interventionist account of downward causation.

2. The Dilemma

What sort of activity is the pursuit of the metaphysics of causation? More narrowly, what sort of knowledge does it aspire to provide us? The title[3] of Mackie's influential 1980 treatise, *The Cement of the Universe: A Study of Causation*, indicates a goal of illuminating at its foundations the connectedness of things in the world. If this illumination is factual, the simplest vehicle for communicating its factual content is through the assertion of a principle of causality. This rather distinctively nineteenth century notion[4] found multiple, largely similar expressions. Here is one example from Flemings' "*Vocabulary*" (1860, p. 78, Fleming's emphasis):

> The belief that every exchange implies a cause, or that every change is produced by the operation of some power, is regarded by some as a primitive belief, and has been denominated by the phrase, the *principle of causality*.

---

[3] We learn from the front matter of Mackie's volume that the expression "Cement of the Universe" is derived from an eighteenth century abstract of Hume's *Treatise*.

[4] A search on Google's Ngram Viewer (https://books.google.com/ngrams/) indicates that the terms "principle of causation" and "principle of causality" seem to have burst with force into the literature in the nineteenth century and enjoyed a strong presence through the century. That presence waned in the twentieth century.

That the factual illumination should come in the form of a principle need not be assumed. The *Cambridge Dictionary of Philosophy* (Audi, 1999) has many entries on causation but, as far as I can see, the terms "principle of causality" and "principle of causation" appear nowhere in it. We cannot even assume that factual illumination is sought. Twentieth and twenty-first century writing on causation is, at least in my reading, quite evasive on the question.

This obscurity of purpose cannot be left unchallenged. Whether factual illumination is provided by the present literature is something that should be decided. To this end, in my (2003) "Causation as Folk Science," I tried to capture in the broadest form what would underlie efforts to provide factual illumination of things in the world. I proposed that such efforts presume (p. 3):

> *Causal fundamentalism*: Nature is governed by cause and effect; and the burden of individual sciences is to find the particular expressions of the general notion in the realm of their specialized subject matter.

A common element in this strain of thinking is the idea that the issues addressed in causal metaphysics come prior to the mundane empirical work of the sciences. That this is the wrong way to proceed was pointed out forcefully by Ronald Fisher, after the coming of quantum theory shook the nineteenth century confidence in determinism. In the first issue of the new journal *Philosophy of Science*, Fisher described what he saw as the failed approach taken to causation in prior centuries (1934, p. 100):[5]

> In the last few centuries a certain uneasy compromise may be observed, between inventing the world *a priori*, and looking to see what it is like. The framework of our thoughts, our preliminary concepts, or basic ideas, have been supposed to be given by philosophy; it was the business of observational science to fit into this framework its missing details, in all confidence that whatever could really be found, had already its place prepared for it in our conceptual framework of ideas about what the world is like.

One might expect this ill-fated approach to have faded away with the new century and its unexpected scientific discoveries. It did not. For example, Hugh Mellor in his treatise, *The Facts of Causation*, wrote (1995, p. 5):

> So while I will accommodate the relevant results of modern physics I will not for example leave it to quantum physics to tell me whether causation can act immediately across spacelike intervals. On the contrary: only when our

---

[5] I thank Brian McLoone for drawing my attention to Fisher's paper.

metaphysics has told us what causation is can we see if physics could reveal unmediated action at a distance (it couldn't).

This insistence sharpens the problem. Metaphysical analysis is providing us something prior to the empirical work of a science—in this case quantum physics. What is the analysis providing? Is it providing us universal facts of causation? Or something else? And, from an empiricist perspective, how can it inform us about quantum processes antecedent to the empirical investigation of quantum processes. To bring the problem into clearer focus, in my 2003 paper, I posed a dilemma for causal fundamentalism (pp. 3–4):

> *Causal fundamentalist's dilemma*: EITHER conforming a science to cause and effect places a restriction on the factual content of a science; OR it does not. In either case, we face problems that defeat the notion of cause as fundamental to science. In the first horn, we must find some restriction on factual content that can be properly applied to all sciences; but no appropriate restriction is forthcoming. In the second horn, since the imposition of the causal framework makes no difference to the factual content of the sciences, it is revealed as an empty honorific.

This early formulation of the dilemma, I now see, oversimplified the second horn. In it, causal analysis is reduced to labeling and does not restrict a science to conform with a prior factual principle of causation. However, unlike the suggestion of the 2003 wording, the labeling need not be empty. It may provide pragmatic benefits to our understanding[6] and practical benefits if it tracks practically useful factual distinctions already present in processes, prior to the labeling.

In the following two sections, I review the prospects for each horn and explain why efforts at accepting the first horn have failed.

3. First Horn: The Failed Quest for Factual Content

The prospects of meeting the requirements of the first horn are poor. To be clear, it is insufficient in this horn for causal metaphysics merely to recite some fact of a connection in the world. It is the case that electric fields produce electric currents if free charges are present. This result and a myriad like it are just the routine results of science. Causal metaphysics pretends to knowledge of causes antecedent to the specifics of such results. It supposes a grasp of causation factually at a quite

---

[6] The term "understanding" is used informally here and below, in a way that is antecedent to the foundational accounts of the term in the recent philosophy of science literature.

general level. What is this general grasp? What in *addition to the individual results of the sciences* has causal metaphysics provided factually?

The short answer is: nothing. That is, the centuries-old history of efforts by causal metaphysicians to deliver factual claims of foundational import has been one of sustained failure. It is a story of proposal and refutation, of confident assertion and apologetic retraction. Here is a sampling of these episodes.

3.1 Early Failures

According to Aristotle we explain the fall of heavy bodies and the rise of light bodies in part through their final causes: they move in order to occupy their natural places. A standard, early example of the repudiation of a causal notion was the seventeenth century elimination of final causes from accounts of the mechanics of bodies. Those accounts were to be based on efficient causes only: the motions of the bodies, their collisions and, after Newton, the forces acting. Part 1, Article 28 of René Descartes' *Principles of Philosophy* (1982, p. 14) was headed with the proposition:

> That we must not examine the final causes of created things, but rather their efficient causes.

As he makes clear in the text of Article 28, Descartes' intent was not to eliminate final causes completely but to remove them to the more elevated sphere of God's inscrutable purposes.[7] This tolerance did not last. Laplace's 1814 *Essai* (1902, p. 3) is quite forthright in denouncing the supposition of final causes for events. They result from "ignorance of the ties which unite such events to the entire system of the universe." Laplace reports triumphantly, "these imaginary causes [including chance] have gradually receded with the widening bounds of knowledge and disappear entirely before sound philosophy."

The central hero of the new era in physics, Isaac Newton, was not immune to failed causal pronouncements. His mechanics provided no mechanism mediating the gravitational attraction of the sun, millions of miles away, upon us here. Nonetheless, he denounced the idea of unmediated action at a distance (Newton, 1761, pp. 25–6):

> [T]hat one body may act upon another at a distance through a vacuum, without the mediation of anything else, by and through which their action and force may be

---

[7] As Osler (1996) has shown, this relocation of final causes from the simple mechanics of bodies to the larger theological sphere was then a common view. Robert Boyle (1688), who coined the term "mechanical philosophy," held a version of it.

conveyed from one to another, is to me so great an absurdity, that I believe no man, who has in philosophical matters a competent faculty of thinking, can ever fall into it.

Yet for all Newton's self-assurance, his theory of gravity soon came to be established as the premier example of an action-at-distance theory. As late as 1901, in his authoritative Teubner Encyclopedia article on gravitation, Zenneck (1901) reported the failure of empirical efforts to see any effect of an intervening medium on the propagation of gravitation (§17, pp. 42–3) and to discern a finite speed of propagation of gravitation (§20, pp. 44–6). Against all efforts to show otherwise empirically, the best account of gravitation remained that it is an instantaneous, unmediated action at a distance.

These are just two of many causal dicta widely accepted in these earlier times, but which do not survive scrutiny. I can continue here only with a brief mention of more of them. In his *Treatise*, Hume (1739, p. 78, his emphasis), for example, reported: "'Tis a general maxim in philosophy that *whatever begins to exist, must have a cause of existence.*" Hume proceeded to attack the maxim from the perspective of his rather severe inductive skepticism. Russell (1912–13, pp. 9–12) in his then inflammatory "On the Notion of Cause," listed five maxims that, he urged, "have played a great part in the history of philosophy." He proceeded to scrutinize and impugn them (Russell's emphasis and quote marks):

(1) "Cause and effect must more or less resemble each other."

(2) "Cause is analogous to volition, since there must be an intelligible *nexus* between cause and effect."

(3) "The cause *compels* the effect in some sense in which the effect does not compel the cause."

(4) "A cause cannot operate when it has ceased to exist, because what has ceased to exist is nothing."

(5) "A cause cannot operate except where it is."

3.2 The Nineteenth Century Dogma of Determinism

In the course of the nineteenth century, causal theorists freed themselves from most of the complications of these maxims, save one. The notion of causation was reduced to one idea, determinism: the present state fixes a unique future state. The causal character of the world inhered in the fact of its determinism. Laplace's 1814 *Essai* gave an early definitive statement of it (1902, p. 4):

We ought then to regard the present state of the universe as the effect of its anterior state and as the cause of the one which is to follow.

Laplace made the determination vivid through his fictitious calculator who could inerrantly calculate the unique future from a full specification of the present. Laplace concluded "for it, nothing would be uncertain and the future, as the past, would be present to its eyes." Probabilities, the subject of Laplace's *Essai*, merely reflected our ignorance of the immense complications of the full determination of things.

John Stuart Mill, perhaps the leading author on methodology in the nineteenth century, based his analysis of causation on this same identification of causation with determinism. He wrote in his *System of Logic* (1882, Bk. III, Ch. V, §2, p. 236):

> The Law of Causation . . . is but the familiar truth, that invariability of succession is found by observation to obtain between every fact in nature and some other fact which has preceded it; independently of all considerations respecting the ultimate mode of production of phenomena, and of every other question regarding the nature of "Things in themselves."

This identification of causation with determinism had the immense advantage of giving a clear and simple account of the nature of causation. It soon became clear that this appeal of simplicity and clarity set the view up for disaster. In the early decades of the twentieth century, quantum theory emerged. The theory contradicted determinism. The best its laws could provide, in the general case, was merely probabilistic assurance of future states, given present states. The quantum theory was rightly recognized as a catastrophic blow to the nineteenth century conception of causation. Friedrich Waismann was a core figure in the logical positivist movement of the early twentieth century. He reflected in dark tones on not just the demise of determinism but the notion, championed by Niels Bohr, that even classical descriptions in space and time fail. In his 1958 lecture, "the Decline and Fall of Causality," Waismann (1959, ch. V, p. 208) lamented that "causality has definitely come to an end: atomic science has penetrated to a depth where an entirely new orientation is called for."

3.3 The Twentieth Century Dogma of Probabilism

This obituary for causality proved premature. Even as Waismann was lecturing, a resuscitated causality was on the rise. Causes no longer determine their effects uniquely, the resuscitation asserted. They only affect their probabilities. This view remains today a dominant view of causality. In his synoptic encyclopedia article, Hitchcock (2021, §1.1) summarizes the new tradition:

> The central idea behind probabilistic theories of causation is that causes change the probability of their effects; an effect may still occur in the absence of a cause or fail to occur in its presence.

There are too many treatments of this notion of causation to survey tractably. Instead, we can sample the literature in Mellor's 1995 *Facts of Causation*. The driving theme throughout is to reassure readers that the loss of determinism does not mean the loss of causation. He bases his case in Chapters 6 and 7 on identifying certain "connotations" of causation: causes are evidence for their effects, causes can explain their effects, and causes are means of bringing about their effects. Mellor's core claim is that these connotations do not require that causation is determinism. Rather they *necessitate* a probabilistic conception of causation (p. 67):

> I shall argue in this chapter and the next that causation's connotations require every cause to raise the chances of its effects.

For "chances" here, we are to understand probabilities (p. 21, Mellor's emphasis):

> The first point I must make about chances is that they are *probabilities*. My other claims about chance may be contentious, but not this one: everyone measures chances by numbers which satisfy the standard calculus of probabilities.

Note that the claim is very strong. Causal connotations "require every cause." No instances of causation escape this theory.

Mellor's conception of the relevant type of probability is correspondingly strong. He insists (p. 75) that "chances must be more than evidential probabilities. ...They measure possibilities... that are properties of facts about the world, and not merely of facts about our knowledge of the world."

Here Mellor has chosen one particular way to implement the probabilistic notion of causation using a physical or objective notion of probabilities. Others working in the subjective tradition of de Finetti may prefer to employ only a subjective notion of probability. For my purposes it is immaterial which particular sense of probability is employed in the account of probabilistic causation. They are all versions of a dangerous dogma:

> *Probabilism*: All indefinitenesses, whether of physical determination of events or of epistemic uncertainty, are always properly represented by probabilities.

Probabilism is presumed throughout Mellor's volume. Chances are identified as probabilities. Evidential support is presumed always probabilistic. This presumption is reflected throughout this literature.

3.4 Problems of Probabilism

The problem of probabilism is not that probabilities never apply. There are numerous cases of probabilistic relations that we might comfortably call causal.

The problem is the insistence that causal connections are necessarily probabilistic; and that this probabilistic conception of causation is universal and exhaustive. That insistence makes probabilism a dogma that fares no better than the dogmas that have come before it. Just as nineteenth century thinkers had an unquestioned but misplaced confidence in determinism, so twentieth century probabilists have an unquestioned, misplaced confidence in probabilism.

The case against this dogmatic version of probabilism has been made at some length in Chapters 10–16 of Norton (2021). The central idea is that probabilities cannot be presumed by default as the way to represent all indefinitenesses. Rather, the applicability of probabilities to some particular instance of indefiniteness must be established by displaying what in the factual background authorizes the probabilities.

One strand of the case against probabilism reviews the range of arguments for the necessity of probabilities. They are, I find, all circular. For the applicability of probabilities to some system is an empirical matter. Thus, any deductive proof of their universal applicability must covertly assume contingent facts logically as strong as or stronger than the assumption of probabilism itself. Once one knows to look for it, it becomes straightforward to identify the presumption of probabilism hidden in each argument's premises.

Another strand of the case identifies systems in which, demonstrably, probabilities cannot be used to represent indefiniteness. These are cases of indeterminism in some physical theory in which the physical theory itself entails indeterminism only, but provides no measure to weight the different possible futures. To represent the indeterminism probabilistically is to add illicitly physical content to the physics applicable. These sorts of systems have been in the philosophy of science literature since Earman's (1986) path-breaking *Primer on Determinism*. Simple examples are collected in Norton (2021, Chs. 13, 15). Most of these cases arise in artificial contexts.

The simplification afforded by the artificial contexts enables quick proofs of the inapplicability of probabilities. Some, however, come from present science. The enduring "measure problem" in present inflationary cosmology is that the theory can provide no unique probability measure over the different non-inflating universes that can be spawned by a larger inflating cosmos. (See Norton, 2021a.)

Familiar background facts that do warrant probabilities can be supplied by a physical theory, such as versions of quantum theory. In the social sciences, probabilities may be introduced by the assumption that individuals of interest in a large population are selected randomly, where the import of "random" is a selection with equal probability.

If there are no authorizing facts but probabilities are still applied, there is a significant risk that results are introduced spuriously as artifacts of an inappropriate representation. For examples, see Norton (2021, Ch. 10, §4) and also Norton (2010). In them, strong conclusions are wrestled implausibly from assumptions that lack factual content sufficiently strong to support them.

In one such case, Van Inwagen has argued that, antecedent to specific knowledge of the world, it is very probable that there is something rather than nothing. There are, he notes, very many ways that things might be, but only one way that things might fail to exist. We do not know which of the ways is the case. We represent that ignorance by distributing probabilities fairly uniformly over all the possibilities. The result is that the overwhelming bulk of probability is assigned to the many ways that things might be. We are to conclude that, with this overwhelmingly high probability, there must be something. The result is clearly spurious. The fallacy resides in assigning probabilities in circumstances of complete ignorance, for those circumstances provide no background facts to authorize the application of probabilities.

This fallacy arises in the literature in many forms. The "doomsday argument" considers a process that we observe has endured for some time t and we are otherwise in total ignorance as to how long T the process persists. The key step in the ensuing argument is to assume that, since we are in complete ignorance as to when the observed time t arose in the overall time T, we say it can arise with equal probability anywhere in the overall process time T. An application of Bayes' theorem then assures us that any smaller value of T is more probable than any larger value of T. There is no background fact warranting the probabilistic representation of this ignorance. The result is merely an artifact of applying probabilistic reasoning illicitly.

If a probabilistic notion of causation is applied to systems indiscriminately without attention to the background facts that authorize the probabilities, there is a significant danger of spurious results. In new work, Wysocki (manuscript) argues that at least some of the cases now routinely dealt with by probabilistic notions of causation can be analyzed without probabilities; and may be better treated that way.

3.5 A Failure of A Priori Science

This concludes a brief review of efforts over the centuries to base a metaphysics of causation in some factual restriction on processes.[8] What we have seen is a sustained record of failure. Time and again, we are offered some pronouncement on the nature of causation as a deep metaphysical truth. New discoveries in science or just closer scrutiny then reveal that the pronouncements are no truths at all, but merely a reflection of the prejudices and dogmas of the moment.

---

[8] This survey does not exhaust the present candidates for a metaphysical necessity of causation. Curie's principle, for example, requires that any symmetry of a cause must be reflected in the effect. Closer examination in Norton (2016), however, shows that the principle exploits a malleability in how we identify causes and effects in some particular system. By careful identifications, we can render the principle true or false in many cases at our whim.

That this is the outcome should, on reflection, be no surprise. The idea that metaphysicians can anticipate the form that all future science must take is doomed to fail. It amounts to an attempt to do science a priori. This is precisely the project that Nature seems to enjoy confounding.

We should not confuse this record of failure with the fallibilism of science. The history of science is replete with cases of theories confidently pronounced and subsequently retracted. The difference is that science has within it the means to correct itself. The major corrections come as empirical investigations advance and reveal more of the world. A priori metaphysics has no corresponding mechanism within it. All it can do is pronounce and then, when the science advances, retract and move on to the next pronouncement.

For example, nineteenth century Newtonians were assured by their physics that the fall of bodies and the motion of planets are deterministic. They presumed this determinism prevailed on all scales, including the very small. That presumption was overturned in the twentieth century when new investigations were able to probe empirically the behavior of particles at the minute atomic and subatomic levels. The revised science now allowed that determinism was preserved for the domain in which the empirical evidence for Newton's mechanics was found: ordinary falling bodies, planetary motions, and the like. The atomic and sub-atomic levels would now be governed by a new, indeterministic quantum physics. This is a story of science evolving as it should by responding to the demands of an expanding body of evidence.

As long as the causal metaphysics merely echoes what the latest empirical science asserts, then the transition is comparable to that of the science. However, that benign circumstance depends on the emptiness of a priori, causal metaphysics: it has no independent grasp on causation prior to the empirical investigations of science. If we suppose that the nineteenth century metaphysical identification of causation with determinism derives from some independent, prior knowledge of causation in the metaphysics, then the diagnosis must be that this prior grasp was mistaken, yet again.

4. The Second Horn: Causal Metaphysics Is Not Factual

The causal metaphysician's dilemma is that we must choose between two horns: EITHER conforming a science to cause and effect places a restriction on the factual content of a science; OR it does not. The discussion of the last section shows the first horn is untenable. As a matter of deductive logic, then, we must accept the second horn. Causal metaphysics adds nothing factual to what we already know by empirical means in our sciences. That is, causal metaphysics has failed to provide novel facts by any means other than through its repetition of what is discovered empirically in scientific investigations.

This result does not deprive causal analysis of its value. Rather it requires us to reconceive its nature. In so far as it is useful, causal metaphysics does not tell us about the world. It tells us about ourselves. There are many ways that this might be so.[9] In my view, the most significant is that causal conceptions can have pragmatic value. We have a naïve conception of simple causal processes derived from our common experiences. When we push (the cause), a body moves (the effect). We treat simple cases like this as a template to be imposed on processes more remote from direct experience to enable us to comprehend them better. The template is imposed by redescribing processes in suggestive causal language, while not adding anything factually to the narrative.

Consider, for example, Ohm's law in simple electrical circuit theory. The voltage drop V in some conductor equals IR, the product of the current I and the resistance R. That is the full content of the law. We are able to use the law better if we attach causal labels to its parts. The voltage difference V is the cause, we might say, that pushes the electric charges through the conductor, producing the effect, I, the current. We have added nothing factual to Ohm's law with this narrative. But it does give us a better grasp on the process: the voltage difference pushes the charges similarly to the way we might push a body. The addition of such new conceptions to the original science by this causal language is described in Norton (2003) as a "folk science."

More abstruse examples are possible. Electrons in atoms in a radiation field, jump up and down among different energy levels, while absorbing and emitting electromagnetic radiation. Einstein's celebrated "A and B coefficients" paper of 1917 summarized the three component processes. (a) In the presence of a radiation field of the right frequency, an electron can, with some probability, absorb energy from the radiation field and jump to a higher level. (b) When at the higher level, it will with some probability emit radiation and drop back to a lower level. (c) This probability is greater in the presence of a radiation field of the right frequency.

We commonly add causal language to this bare but factually adequate account. In (a), the electron is said to be induced—caused—to jump to a higher energy level by the radiation field. It does so in much the same way as we may induce a guest to a second helping of pie. In (b), the electron is drawn back to its natural, lowest energy state, which we conceive as the applicable final cause. Finally, in (c), the

---

[9] Here many find Kantian approaches to causation helpful. In them, causal notions arise of necessity through the way that we interact with the world. These are attempts to secure a priori necessities in another way. However, their prospects are no better than the other a priori dogmas of causation sketched in the last section. We simply are not good at intuiting a priori how things might unfold in the world. For all his prodigious mental powers, Kant's thinking was still confined by the prejudices of his era. Contrary to the certainties of his time, the geometry of space does not have to be Euclidean. We have no reason to expect present day Kantian approaches to fare any better, in so far as they arrive at facts by a priori means.

electron is stimulated by the field to jump down to the lower energy state,[10] in much the same way that the cat might be stimulated by a tickle to jump down from the sofa. Once again, adding the causal language "stimulated" has not added factually to the description of the process, but it has enhanced our grasp of it.

Examples like this show how readily we can add causal language to some process without adding factually to it. As far as I can see, there is no limit to this process. We can, if we wish, always find a way to add causal descriptions to any science. Any process can be conformed to some conception of cause and effect, without placing a factual restriction on the process.

5. Causation Is Not Factual Discovery, but Convenient Definition

The analysis of the causal fundamentalist's dilemma leads to this result: the identification of causal processes in the world is not one of factual discovery, but of the application of convenient definitions. When we consider some process, there is no fact of the matter as to whether this particular process is causal or not. Whether it is or not is just a matter of how we decide to use the term "causal." This is what we have learned from the fragility of the many causal dogmas. What were yesterday's truths of causation are today's errors. The indeterminism of quantum processes was received in the 1920s with alarm. "[C]ausality has definitely come to an end." This was Waismann's appraisal, as reported above. A century later it is standard to think otherwise. That the connections of quantum theory are merely probabilistic places them comfortably within routine causal analysis. Nothing factual has changed. The only difference is our choice of which definitions are congenial.

My fear is that readers, familiar with Humean skepticism, will misread these last claims as just another version of Hume's causal skepticism. He was averse to such notions as "efficacy, agency, power, force, energy, necessity, connexion and productive quality" (Hume, 1739, p. 157). Where others saw them, all Hume saw was the habit of mind of expecting past regular connections to continue.

My view is not a Humean skepticist one. I fully accept that things connect in a myriad of ways; that science has revealed all manner of hidden powers and productive qualities; and that here science enjoys uncommon success. What I deny is that causal metaphysics has anything factual to add to these successes. Tides on the earth's oceans correlate with the positions of the moon and sun. Hume regarded this merely as a constant conjunction. I do not. The tides are raised by the gravitational action of the sun and moon. Call that action "causal" if

---

[10] This stimulated emission is the process of a LASER = Light Amplification through Stimulated Emission of Radiation.

you want. In so doing, nothing factual is added. All I have learned is how you conceive of the process and where you find causal labels apt.

Under this deflationary view of causation, philosophical theories of causation amount to declarations about how the theorist will use a term. For example, regularity theorists announce that the terms "cause" and "effect" can be used whenever certain sorts of regularities manifest between events of specified types. Those advocating a counterfactual theory of causation augment these occurrent regularities with further conditions concerning what regularities might, counter-factually, have occurred but did not. In the Salmon–Dowe process theory of causation (Dowe, 2000), a process is causal if it transmits a conserved quantity. They are the attaching of labels to the behaviors of things in the world as already discovered empirically by the sciences. While these labels may be pragmatically useful, none of them and the other theories like them report a factual discovery about things in the world, made antecedently to empirical investigation. There is no compulsion that the world should present us with behaviors of the requisite type, beyond that provided by the evidence unearthed by the sciences. We have powerful evidence that processes in the world convey the conserved quantity energy and, to the extent that this is true, the Salmon–Dowe theory will be able to identify processes to which its causal label can be attached. There is no a priori necessity that all processes must be such. Recent science has at least toyed with the idea of processes that we might like to call causal but which do not pass energy. In traditional accounts of quantum theory, the mere observation of a quantum system can trigger its collapse without the process of observation exchanging energy with the system. Other accounts of causation face more serious challenges. The enduring difficulty of counterfactual theories of causation is that there is no agreement on precisely which formulation is the right one, if the relationship is to be identified as causal.[11]

6. The Manipulability or Interventionist Conception of Causation

Designations of causality are mere matters of definition. That fact, however, does not make the designations worthless. Namings, even arbitrary ones, can have great utility. Without names for streets, towns, cities, and nations, routine travel would be greatly inconvenienced. So far, above, I have suggested that causal naming can enhance understanding. That is a foundationally fragile benefit, for understanding

---

[11] For a recent survey, see Menzies and Beebee (2020). Frisch (2009a, b) has argued for a causality principle in physics in which the effect can never precede the cause. One of the criticisms of the proposal (Norton, 2009) is that it fails to identify what counts as a cause and an effect and to distinguish which is which.

is connected to the idiosyncrasies of the individual. Probabilistic accounts of causation offer some a sense of understanding since now causal relations are expressed in a precise, quantitative theory, the probability calculus. For me, however, the situation is reversed. I find handing over determination of these relations to a computational tool obscures my understanding. When I do the calculation, I can see that conditionalizing on some factor increases the probability of some other factor. However, I need to look behind the probabilities to understand why this increase obtains.

The manipulability or interventionist conception of causation is free of these complications. It identifies a process as causal when our intervention or manipulation of what we will designate the cause results in a corresponding change in what we will designate as the effect. Knowing which processes are causal in this sense has great pragmatic value. The obvious examples are in medicine. We learn that high blood pressure causes strokes and high blood cholesterol levels cause heart disease in this sense. We then know something pragmatically useful: that intervening on high blood pressure and on high cholesterol levels with suitable medications or changes in diet and lifestyle can ameliorate the unwanted effects.

There are many other prosaic examples outside medicine. Water and air cause iron to rust. By intervening to preclude the contact of one or both of water and air with the iron, we can prevent the effect of the rusting of the iron. There are more exotic examples. We noted above that an excited atom can be stimulated to emit radiation if it is immersed in a radiation bath of the right frequency. The manipulation of the surrounding field causes an effect, the emission of radiation. This is how we build LASERs.

The foremost proponent of this approach to causation is Woodward, whose (2003) work is definitive. A version of the account, adapted to variables governed by deterministic relations, is given in a later work (2021, §2) as:

(M)   Where X and Y are variables, X causes Y iff there are some possible interventions that would change the value of X and if were such intervention to occur, a regular change in the value of Y would occur.

The most common setting, however, for this interventionist account is sets of variables that are nodes in an acyclic, directed graph of relationships, where a probability measure is defined over the variables. The identification of the causal relations then requires some elaborate lawyering. For what is termed "actual causation," the best developed condition in Woodward (2003, p. 84) is the condition AC*. While most of its details will not be relevant to the discussion here and will not be elaborated, the definition is reproduced to give a sense of the care and richness of the specification. The notion is defined over the variables mentioned in the formulation in upper-case letters and their values in the corresponding lower-case letters:

(AC*1)   The actual value of $X = x$ and the actual value of $Y = y$.

(AC*2)   For each directed path P from X to Y, fix by interventions all direct causes Z_i of Y that do not lie along P at some combination of values within their redundancy range. Then determine whether, for each path from X to Y and for each possible combination of values for the direct causes Z_i of Y that are not on this route and that are in the redundancy range of Z_i, whether there is an intervention on X that will change the value of Y. (AC* 2) is satisfied if the answer to this question is "yes" for at least one route and possible combination of values within the redundancy range of the Z_i.

$X = x$ will be an actual cause of $Y = y$ if and only if (AC*1) and (AC*2) are satisfied

What will matter for the further discussion here is that this definition and others like it is incomplete without a definition of "intervention." That is provided as IV (p. 98):

(IV)

I1. I causes X.

I2. I acts as a switch for all the other variables that cause X. That is, certain values of I are such that when I attains those values, X ceases to depend on the values of other variables that cause X and instead depends only on the value taken by I.

I3. Any directed path from I to Y goes through X. That is, I does not directly cause Y and is not a cause of any causes of Y that are distinct from X except, of course, for those causes of Y, if any, that are built into the I-X-Y connection itself; that is, except for (a) any causes of Y that are effects of X (i.e., variables that are causally between X and Y) and (b) any causes of Y that are between I and X and have no effect on Y independently of X.

I4. I is (statistically) independent of any variable Z that causes Y and that is on a directed path that does not go through X.

The essential notion here is that of a "switch." Other similar accounts employ a related notion. Pearl (2009, p. 70) employs a "do" operator that models an intervention as the setting of the value of some variable to a stipulated value. It necessarily breaks some dependencies, while leaving as many others unaltered as possible.

   In my view, the interventionist account is probabilistic only by an accident of the context in which it is commonly implemented. In my version of interventionism, what is essential is the notion of an intervention or manipulation. Thus, Woodward's later deterministic version "(M)" above gives a clearer presentation of the core idea. Pearl's "do" operator shows most directly that the probability distribution prevailing over the variables is inadequate to capture the notion of intervention. The "do" is implemented by breaking some relations through means outside the probability distribution. Woodward's condition "(IV)" seeks to implement the notion within the existing probabilistic setting. We shall see below that this leads to one of the enduring problems of the interventionist account.

This interventionist account should be contrasted with other accounts that are essentially probabilistic. Unlike the interventionist account, such accounts cannot be formulated without the probability distribution. In them, at their simplest, a causal relation is identified merely when conditionalizing on what will be designated a cause within the distribution raises the probability of what will be designated the effect. Or, in more complicated cases, such as Reichenbach's (1971, p. 157) common cause principle, conditionalizing on a common cause eliminates probabilistic dependence among the effects. Unlike the interventionist account, such theories cannot be formulated without probability measures.

7. Solving Problems for the Interventionist Account

Two problems have routinely troubled the manipulability or interventionist account: a concern over its limited scope and a threat of circularity. These are long-standing problems and were already addressed at some length in an early development of the interventionist account by Menzies and Price (1993).

7.1 The Problem of Limited Scope

The interventionist account can only identify a causal process if we can intervene on one of its variables. There are numerous processes that we might like to call causal in which this is not possible. The earth's magnetic field derives from electric currents associated with the convection of molten metals inside the earth. Or, to use an example from Menzies and Price (1993, §5), earthquakes arise from a shift in the tectonic plates of the earth's crust. We might like to say that the convection currents cause the magnetic field and the plate shifts cause the earthquakes. However, we have no practical means of intervening or manipulating these causes.

It is tempting to extend the interventionist account by analogy. The process producing the earth's magnetic field is analogous to the production of a magnetic field by an electric current in a laboratory solenoid. That electric current can be manipulated by a simple switch in the circuit and is the cause, in the interventionist sense, of the magnetic field. We might reason similarly in the case of earthquakes by considering how we might manipulate a laboratory scale model of the tectonic plates.

Woodward (2016, §2) has already raised concerns with this extension. How are we assured that the target system and the model are sufficiently analogous for the causal claim to carry over? That would seem to require some alternative conception of cause that has already been found to apply to the target system. Worse, we can proceed to more extreme examples in which the analogy is implausible. Laplace, we saw above, declared the universe at one time to be the effect of its earlier state and the cause of its future state. Thus, we might now conceive the unimaginably hot early universe of modern cosmology as the cause of our present universal state. However, barring fictional fantasies, manipulations of this early state are implausible even in analogy.

These are practical obstacles to extending the interventionist conception more broadly. My view is that there simply is no reason to extend it and good reason not to. We might like to think that the early universe is the precursor of its present state; or that internal currents produce the earth's magnetic field. However, nothing factually is added to the existing cosmological theory and to the existing electrodynamics by appending the appellation "cause." To feel a pressure to extend the interventionist conception to cases where no intervention is possible is to fall prey to defective causal metaphysics. We would fall into the illusion that there is some deeper fact of causation in the world that goes beyond the explicit content of the sciences. We would be accepting a mistaken view: that these are just two examples of an enormous repository of canonical cases of this true sense of causation that extend beyond cases in which interventions are possible; and that any account of causation must somehow accommodate all these cases.

On the contrary, the strength of the interventionist account is that it gives us a pragmatic reason to designate certain processes as causal. They are processes that we can deploy to our advantage. To seek to apply the theory to cases in which such pragmatically useful manipulation is not possible is to undermine the very strength of the theory. There is no point in trying to find an interventionist apology for these cases. For they have no special claim on us in the first place. Laplace found it apt to label the past as the cause of the present without that designation adding anything to the physical theory that relates the past to the present. Seeking an interventionist interpretation of the relation has no pragmatic value. It dilutes and confuses the interventionist theory by depriving it of exactly what makes it work: that causes are identified as things that can be manipulated.

This last analysis also responds to another objection discussed in Menzies and Price (1993, §6). Is the interventionist conception unacceptably anthropocentric? The answer here is that the account, or at least my version of it, is essentially and irremediably anthropocentric. It is all about what we can manipulate in the world. That is what makes the theory useful and it needs to offer no apologies for it.

7.2 The Threat of Circularity

An enduring issue with the interventionist theory is the accusation that the theory is circular and thereby undermined. The concern appeared early in Menzies and Price (1993, §4) and still required discussion in Woodward's Stanford Encyclopedia article (Woodward, 2016, §6). The difficulty is that an intervention is itself routinely understood to be a causal process. Thus, any attempt to define causal processes in terms of interventions presumes the very thing that is to be defined. The difficulty is quite clear in Woodward's (2003, p. 98) definition IV above of what it is for I to intervene on X. Its first clause is just "I1. I causes X."

While the literature has many responses, it seems to me that the simplest follows directly from the essential anthropocentrism of the conception. My austere version of interventionism can be reduced to two slogans:

A manipulation or intervention need not be a cause; and at least some are not.

A causal process is the propagation of an intervention or manipulation through a network of dependencies.

That is, we take the notion of an intervention, at least in some cases, as a primitive, deriving from our routine understanding of human actions. Once we have introduced an intervention in this primitive sense into the analysis, we can, if applicable, use the causal processes that ensue as interventions within further causal process. This expands the scope of the interventionist theory. We can set up chains of interventions. We intervene on this variable to manipulate another variable; and that manipulation is an intervention that leads to manipulation of a third variable; and so on.

An intervention, in this primitive sense, might arise when we intervene on the electrical conductivity of some component in an electrical circuit by throwing the switch. The propagation of that intervention leads to the light illuminating. The propagation, but not the intervention, is causal. The temptation is to try to apply the interventionist theory to the human action of throwing the switch. Might we say that my willing my hand to move intervened in my nervous system in such a way as to lead my hand to throw the switch? Since I have no special expertise in psychology and neuroscience, I have no idea if something like this can be offered responsibly as an account of my action. However, I am confident that any such account will merely push back the point at which interventions as a primitive are needed in the theory.

Attempts to avoid a primitive notion of intervention have not fared well. We might temporarily manage if there is a cascade of causal processes such that each cause in the cascade serves as an intervention for the next. However, we eventually run into the problem of characterizing the first intervention in the cascade. I read

Woodward's definition IV as successfully characterizing interventions that arise as intermediaries in a causal cascade. However, the intervention at the outset of the cascade eludes the definition. There is no earlier intervention to allow it to be identified as causal. Pearl's "do" operator is unapologetically a primitive that lies outside the existing probabilistic relations prevailing over the variables in his directed graphs.

To accept the notion that some interventions are an ineliminably human, primitive notion, as my version does, is no compromise. It is to accept that the theory derives all its value from the pragmatic benefits it affords humans. That the theory depends on a human notion, primitive and irreducible in the context of the theory, is simply a core element of the theory. To try to explicate further this human notion is to undertake a project that belongs elsewhere in science, perhaps in psychology and neuroscience. To regard explication of the notion as outside the scope of the theory is not to offer it as a mystery. It is merely to say that, if it is treated as a primitive, it is a notion well enough understood for the purposes of the interventionist theory; and that further investigation of it belongs elsewhere.

8. Downward Causation

The failure of determinism in the early twentieth century was a disaster for causal metaphysics. My impression had been that causal metaphysicians were sufficiently chagrined by it that they subsequently avoided strong factual claims about causation in favor of accounts that could be read as definitions. That meant that a factual principle of causality or other causal maxims of comparable factual import started to fade from the literature. Such assertions had not fared well in the past. Why risk more of them?

While such prudent hesitation may be true for the mainstream of metaphysicians constructing positive accounts of causation, it turns out not to be the case in adjacent literatures. A notable example concerns the recent debate over "downward causation." An enduring approach by critics of downward causation is to argue that downward causation violates one or more supposedly unimpeachable maxims of causation. However, a closer examination of the maxims shows that none of them are sustainable as factual restrictions. They are not expressions of some factual principle of causality that has for so long eluded causal metaphysicians. In keeping with the general approach of this chapter, my view is that whether downward causation occurs is not a factual matter. Rather its occurrence depends entirely on the definition of causation one finds apt. In this section, I will argue that interventionist accounts of causation do support downward causation. In the following section, I will argue for the failure of attempts to impugn this interventionist analysis by means of the causal maxims.

8.1 Downward Causation

Downward causation arises when causation acts from a higher level of description of some system to a lower level. Specifying it requires an account of levels. A common example is that thermodynamics provides a higher-level account of thermal systems; and the statistical physics of molecules and radiation modes provides a lower-level account. Another is a motivating example for this literature. It concerns mental causation. The higher-level relations are among mental states and actions, such as willings of some behavior; and the lower level concerns neuronal activity that implements the action. There is much more to say than can be said here. In their introduction to an eighteen-chapter collection of papers on downward causation, Paoletti and Orilia (2017, p. 9) draw a list of roughly thirty proposed examples of downward causation from the volume. They derive from physics, chemistry, biology, neurosciences, psychology, and sociology. Ellis (2009, p. 64) identifies a hierarchy of eight levels that rise from particle physics to sociology/economics/politics. Each level harbors its own phenomenological theories and each provides the possibility of top-down causation. "Top-down causation is ubiquitous in physics, chemistry, and biology," he concludes (p. 66). As the size of the Paoletti and Orilia volume and other related works suggests, the literature on downward causation is large. My account here can only explore a small portion of it that pertains specifically to issues of this chapter.

While the debate over downward causation is energetic and long-lived, from the perspective of the present analysis, the debate depends on its participants employing different definitions of causation; or even just vague definitions of causation. That definitions are so central explains a striking feature of the debate: the factual relations prevailing in disputed cases are often fully known, so that a determination of whether downward causation is present or not adds nothing to our factual knowledge. That is how debates dependent on different definitions manifest. In some cases, however, the prevailing facts are not fully known. The prime example is of mental causation and its foremost exponent is Jaegwon Kim (whose analysis is recounted in the following section). Do our wishes cause our neurons to trigger? What is striking in Kim's analysis is just how irrelevant are detailed factual matters concerning the relationship of mental states and neural states. It is generally supposed that the mental states supervene on the neural states and, merely by reflecting on this supervenience, we can determine whether a causal relation obtains. Notably, a decision in either direction seems to add nothing to our knowledge of the factual processes that connect wishes and neural states.

8.2 The Interventionist Account: The Evaporation of Water

Whether downward causation occurs is, according to the perspective of this chapter, simply a matter of how one defines causation. Depending on the definition one finds apt, it may or may not occur. If, for example, one favors a Salmon--Dowe conserved quantity definition of causation, then likely downward causation is precluded. Exchanges of energy and momentum are most plausibly restricted to one level. This is one objection Gillett (2017, pp. 250–1) raises against downward causation. An interventionist account, however, seems quite amenable to downward causation. That downward causation can be understood in the interventionist conception has been defended by Woodward, most recently in Woodward (2021), and also by Kistler.[12] Ensuring that plausible candidates for downward causation fit with the precise details of the interventionist account has required some elaborate argumentation and delicate adjustments that can be found in their papers. My concern here is not with the details of these adjustments since their success seems assured by the existence of simple examples that conform well with the interventionist conception.

It is easy to see through these examples that the general interventionist idea is quite hospitable to plausible candidates for downward causation. An example I find striking illustrates how this works. The example arises in thermal physics. Warm water evaporates faster than cold water. We can speed up the evaporation of water by intervening on its temperature. We may bring a dish of water into a warmer room; or we may gently warm it on a hot plate; or we may warm it by exposing it to sunlight. These interventions are on the variable of temperature that resides within the higher thermodynamic level of description. On the lower molecular level, evaporation occurs when individual water molecules are energized enough thermally to break free from the forces that hold them in the bulk liquid state. Increasing the temperature of the water increases the speeds of the water molecules and, with it, the probability that individual water molecules exceed this escape threshold. This probability is the lower-level variable affected by manipulation of the higher-level variable, the water temperature.

Two aspects of this instance of downward causation are notable. First, examination of the relation is not just qualitatively useful but also gives useful quantitative results. What follows is a *very rough and ready* illustration. The energies of water molecules are Boltzmann distributed. That is, the probability that a molecule has energy E is proportional to the Boltzmann factor, $\exp(-E/kT)$, where k is Boltzmann's constant. If we extract from this distribution just the kinetic energy associated with the vertical component of velocity v of each molecule (toward the water surface), we can approximate well enough the probability density over the vertical component of the velocity v of molecules of mass m as

---

[12] The proposal has engendered a small debate in the literature. The course of the debate is summarized in Woodward (2021) and Kistler (2017). Ellis' (2009) account also employs interventionist notions, although he does not label them as such. He writes (p. 66): "How do you demonstrate top-down causation? You show that a change in high-level variables results in a demonstrable change in lower-level variables in a reliable way, after you have altered the high-level variable."

$$f(v) = \left(\frac{m}{2\pi kT}\right)^{1/2} exp\left(-\frac{mv^2}{2kT}\right)$$

A first crude model is that a water molecule escapes the surface when its vertical speed v exceeds some value V. The probability that a water molecule exceeds this vertical speed in water of temperature T is

$$Pr(T) = \int_{v=V}^{\infty} \left(\frac{m}{2\pi kT}\right)^{1/2} exp\left(-\frac{mv^2}{2kT}\right) dv = \int_{x=X}^{\infty} \left(\frac{1}{2\pi}\right)^{1/2} exp(-x^2/2) dx$$

where $x = (m/kT)^{1/2}v$ and $X = (m/kT)^{1/2}V$. This integral extends over an exponentially decaying tail, so that most of the probability will be massed close to the lower bound, X. Hence, as a rough approximation, we can assess the effect of varying temperature through the proportionality:

$$Pr(T) \sim exp(-X^2/2) = exp(-mV^2/2kT)$$

Taking logarithms, the temperature dependency of the probability can be expressed as

$$\log Pr(T) \sim \text{const.} - \text{const.}/T$$

Since the rate of evaporation is determined by this probability, we find a "$-1/T$" increase of the rate with temperature.

It is striking that this relation agrees with a corresponding result located entirely within the thermodynamic level. At this level, the rate of evaporation is determined by the vapor pressure P of the liquid. The Clausius–Clapeyron equation relates this vapor pressure P to the system temperature T. Under suitable idealizing assumptions, it reduces to[13]

$$\log P \sim \text{const.} - \text{const.}/T$$

The agreement of the rough molecular calculation with this result is actually too good considering the approximations employed. However, all that matters for our purposes is that this example of downward causation supplies useful quantitative results that could be refined with more realistic assumptions.

---

[13] The Clausius–Clapeyron equation is $dP/dT = L/(T\triangle v)$, where L is the molar specific heat of vaporization and $\triangle v$ is the molar specific volume change on evaporation. We assume that L is constant over the temperature range of interest and that the vapor phase is an ideal gas, so that $\triangle v = RT/P$, where R is the ideal gas constant and the small liquid volume is neglected. The Clausius–Clapeyron equation becomes the differential equation $dP/dT = LP/RT^2$ and its integral solution is the "$-1/T$" dependency indicated.

The second aspect pertains to a common rejoinder to examples of downward causation like this. There is, we are assured, necessarily a causal relation that resides fully within the lower level; and this renders the downward causation redundant. If the water is warmed on a hot plate, the relation would involve the thermally excited energy of molecules of the hot plate, and how that energy is communicated to the water vessel and then to the water. If the water is warmed radiatively, however, the relation would pertain to the energetic coupling of radiation modes in incident sunlight with degrees of freedom of the individual molecules.

No doubt, some definition of cause would apply to these lower-level processes. For example, they involve the transfer of energy and momentum in conformity with the Salmon–Dowe account. However, identifying these relations as causal would have little utility. They do not aid in understanding but merely bury us in masses of superfluous molecular-scale details. They do not enhance our ability to affect the rate of evaporation. That capacity is already provided by the relation of downward cause: to increase the rate of evaporation, increase the temperature by whatever means you like. Indeed, if anyone were to attempt the fully molecular account, they would likely end up noting that conductive or radiative heating merely serves to increase the parameter T in the Boltzmann distribution. In effect, in the essential part, they would end up agreeing with the downward causal account. It would be in a less informative way since the T in the account would merely be a parameter divorced from a higher-level, thermo-dynamic interpretation.

That is, in terms of both understanding and pragmatic application, the downward causal relation is superior to the one instantiated solely at the lower level.

9. Causal Maxims Are Reawakened

Here I will review attempts to impugn downward causation by the suggestion that it contradicts supposedly secure maxims about causation. These attempts fail, I will argue, because of the dubious status of the causal maxims. To summarize what we will see below, there are two difficulties with the maxims.

The first difficulty is that the viability of a causal maxims depends sensitively on the sense of causation employed. A maxim that works for one sense may not for another. Since the Salmon–Dowe approach requires the passage of a conserved quantity from a cause to an effect, it is inherently asymmetric. However, the causal relations in the interventionist account are between variables and thus can go in either direction. Again, we may find that causal closure or completeness fail in quantum mechanics if our conception of causation is deterministic. It may succeed, however, if we posit a probabilistic conception. The result is that no single maxim can assure a blanket preclusion of downward causation. A causal maxim adapted to one sense of causation may not conform with another. This multiplicity of notions of causation is central to Ellis' (2009) identification of five different types of top-down causation that arise according to the context. In reflecting on them, he writes (p. 64, his emphasis):

> There could be others, but I claim that these can all be regarded as well established. In brief: *there are other forms of causation than those encompassed by physics and physical chemistry.* A full scientific view of the world must recognize this fact, or else it will ignore important aspects of causation in the real world, and so will give a causally incomplete view of things...

The second difficulty is more serious. These causal maxims are deployed by critics of downward causation as deep truths, not mere matters of comfortable definition. Thus, they need to be secured by appropriate empirical investigation. Yet these causal maxims are most commonly just declared as self-evident truths or perhaps just what everyone thinks. This is a fallacy of "argumentum ad populum" and not a secure basis for maxims that are to control a major, foundational debate. There are cases, we will see, in which a factual basis is identified. Commonly, the factual basis resides in carefully chosen examples. Yet, as we shall see, we can find other examples that contradict the maxim. More curiously, such identifications are often accompanied by the concession that the maxim has been disputed and has known counterexamples.

The result is that efforts to use causal maxims as a means of impeaching downward causation fail; and, in particular, they fail as objections to the interventionist account of downward causation.

9.1 Causal Exclusion and Causal Closure

Perhaps the best-known arguments against downward causation are offered by Jaegwon Kim in several papers and his two works (1998, 2005). They arise in the context of the philosophy of mind. His "supervenience argument" against downward causation is based on two maxims. The first is the "causal exclusion principle" (2005, p. 42):

> *Exclusion.* No single event can have more than one sufficient cause occurring at any given time—unless it is a genuine case of causal overdetermination.

The second is the "causal closure principle" (2005, p. 43):

> *Closure.* If a physical event has a cause that occurs at t, it has a physical cause that occurs at t.[footnote]

These principles are applied to the two levels in the philosophy of mind: the mental and the physical. If some event at the physical level has a cause, then, by *Closure* it has a physical cause. We may try to offer a mental cause as well. *Exclusion* precludes this mental cause and with it the possibility of downward causation. Kim (2005, p. 44, his emphasis) concludes, "*the assumptions of causal exclusion and lower-level causal closure disallow downward causation.*"

Both these principles are inadequate from the perspectives of this chapter. Both proceed without drawing anything of substance from the empirical investigation of psychology and neuroscience. Hence it is hard to see how these principles can be securely based. Both proceed without giving a clear definition of what is meant by the term "cause." Without it, it is hard to know precisely what they say. We might accept that the ontology of the lower level is exhaustive. That is, in this context, we might accept that all mental activity simply is neuronal. However, that view of the ontology is not sufficient to establish the causal closure principle. The notion of cause at issue must also be specified and then it must be shown that the closure principle applies to it. Might it be that the causes of an effective analysis are only properly expressible at some higher level so that there is no description of the cause in the lower level? Ellis (2009, p. 66) argues that just this does happen since, in his view, top-down causation depends on higher-level context variables:

> Altering the high-level context alters lower-level actions; this is what identifies the effect as top-down causation. In such cases the high-level context variables are not describable in lower-level terms, and this is what identifies them as context variables.

These ambiguities make the principles fragile. Kim's footnote to the above statement of *Closure* illustrates it. The footnote calls the reader's attention to two further sources for discussion. One is Lowe (2003) who points out that quantum theory prevails at lower levels. Under a deterministic conception of causation, probabilistic quantum events such as the death of Schrödinger's cat, have a cause but, he argues, not a sufficient cause. He continues to argue that replacing the deterministic notion of causation with a probabilistic one does not save the analysis. We are assured that "we have every reason to suppose that (CCP) [the causal closure principle] is empirically false, its falsity being testified to by the empirical data which support the principles of modern quantum physics" (p. 142) and finally "I am firmly convinced that (CCP) is false" (p. 145).

It is perhaps germane to add that this supervenience argument does not succeed in impugning most of the cases of downward causation proposed in the present literature. If it works at all, it would only apply to cases of downward causation such as would appear in a dualist theory of mind or other theories like it with a dualist ontology. Then the downward causation is from a mental world to the physical world that is distinct from it.

Aside from these dualist cases and others functionally similar to them, Kim's framework actually affirms downward causation for the common examples in the literature. This affirmation arises because of the way Kim implements the causal exclusion principle. The principle is developed in an earlier paper, Kim (1989). We are asked to consider two explanations for some event (p. 89):

Explanation A cites C as a cause of E

Explanation B cites C* as a cause of E

We are then taken through five cases. The first four are: C and C* are the same; C is distinct from C* but reducible to it; neither C nor C* are individually sufficient as a cause; and C and C* are different links in the causal chain that leads to E. The causal exclusion principle applies to none of these. For none provide independent causal explanations. The causal exclusion principle applies only to the fifth case in which C and C* are distinct and individually sufficient as causes of E. This fifth case includes dualist ontologies, such as a dualist theory of mind.

These first four non-dualist cases are the cases implemented in common claims of downward causation. There we typically have a higher-level description of a cause and lower-level description of a cause. They are commonly the same thing, just described at different levels; or something close to it along the lines of Kim's first four cases above. At the higher level of description, transferring heat to water has the effect of increasing its rate of evaporation. At the lower level, energizing the individual water molecules has the effect of increasing the probability that they escape the liquid phase and evaporate. These two causes—transferring heat and energizing molecules—are just the same thing described at different levels. In such cases, if the thing identified at the lower level is a cause of the effect, then *necessarily* the thing identified at the higher level is also a cause of the effect. It has to be that way since they are the same things, just described at different levels.

We have the curious result that, if we accept Kim's maxims and his understanding of them, his work provides a vindication of downward causation in most of the cases considered in the Paoletti and Orilia (2017) volume.

9.2 Whole to Part Causation, Synchronic Causation

A different causal maxim that is used to criticize downward causation pertains to wholes and parts. Woodward (2021, p. 12) reports:

A very common criticism of the idea of downward causation is that this requires that "wholes" act downward on their "parts" and that the relation between a whole and its parts cannot be a causal relation of any kind.

He identifies Craver and Bechtel (2007) as an instance of it. In their work, the preclusion is justified by dual considerations: that wholes and parts are not sufficiently distinct: and that whole to part causation is synchronous. That is, it violates a second causal maxim that prohibits instantaneous causal actions. Their initial formulation is interesting for present purposes (2007, p. 551):

> Many common assumptions about the nature of causation preclude the possibility of causal relations between parts (components) and wholes (mechanisms). To start with an especially clear example, consider the view that all causation involves transmitting something such as a mark (Salmon 1984) or a conserved quantity (Dowe 2000) from one event, object, or process to another.

The reference to "common assumptions" reflects the concern above of an argumentum ad populum. The attachment to particular accounts also reflects the fragility of the maxim. For what holds in one account may not hold in another. Attached to this maxim is the second maxim that precludes synchronic causation—that is, causation in which the cause and its effect happen at the same time. Their concern is that we might have two systems each instantly causing the other to acquire some causally efficacious property. They continue (p. 553):

> To avoid this problem, one might assume that causal transactions across levels take time: the effects of changes to a component alter the behavior of the mechanism as a whole at some later time, and vice versa.

We see a similar recourse to popular ("generic") views in Gillett's suppositions that could be used to preclude whole to part causation (2017, p. 250):

> I simply use generic features of the causal relation to assess the claim that FDR [Fundamental Downward Relation] is downward causation in such cases. I therefore assume causation to be a relation holding between wholly distinct individuals that is temporally extended and where its *relata* are usually at different locations.

Woodward's (2021, p. 8) defense of the interventionist account against these objections seems to me to be decisive:

> Scientifically plausible examples of downward causation do not involve wholes acting on parts but rather involve variables (as all causal relations do) and these need not stand in part/whole relationships, even when entities of which they are predicated do.

My concern here, however, is the cogency of the maxims that prohibit whole to part causation and synchronic causation. They do perhaps fit with the examples Craver and Bechtel favor. However, they do not fit with others. Laplace, we saw above, regarded the present state of the universe as a cause of its future state. Presumably the future state includes its parts. Thus, the totality of the position and velocities of the sun and other solar system bodies today determines their whole configuration tomorrow. Within that whole will be a part, such as a particular, localized event. It might be an eclipse tomorrow. Thus, today's state of the whole is the cause, in Laplace's sense, of that part of tomorrow's whole.

This last whole to part relation is not synchronous. Today's whole causes something in tomorrow's part. Yet that a causal action is synchronous has historically not precluded it from serious consideration. For hundreds of years after Newton, a paradigm of causation was the gravitational action exerted among bodies. In it, the Newtonian forces acting on a body are the cause of the effect of the body's acceleration. Since Newtonian gravitational forces propagate instantly, the cause and its effect are synchronous. In the early nineteenth century, Laplace tried to place a bound on the speed of propagation of gravity and found such a high speed that he concluded it could be assumed infinite.[14]

If gravitational actions are causal, we may take this example as a case of whole to part, synchronous causation. The cause of a celestial body's acceleration is the entirety of the gravitational action of all the bodies in the universe. If we think in terms of fields, the force on the body is due to the prevailing gravitational field, which is in turn determined through instantaneous action by the masses and positions of all the bodies in the universe.[15]

9.3 The Asymmetry of Causation

Two related maxims concerning causation have also been applied in the literature critical of downward causation. They are:

> The causal relation is asymmetric: causes act on their effects, but effects do not act on their causes.

---

[14] Laplace (1805, p. 526): "one finds [from the secular motion of the moon] that the speed of the gravitational fluid is about seven million times greater than that of light...one may suppose for the gravitational fluid a speed at least a hundred million times greater than that of light....Therefore, surveyors can, as they have done so far, assume that this speed is infinite."

[15] A loophole here is that this entirety may exclude the body in question. Since Newtonian theory is linear, altering the mass of the body does not affect the force acting on it. However, altering its mass will alter the effect described, its acceleration, all else held constant. If this is still a concern, we can consider a non-linear Newtonian-style gravitation theory in which the body's own "self-field" contributes to the acceleration. Such self-field contributions arise in the classical theory of accelerating electric charges.

The causal relation is time ordered: causes precede their effects.

The difficulty with both maxims is that they have been challenged extensively in the present literature. The challenges are so pervasive that the favorable reporting of the maxims commonly include the concession that they are challenged.

Here is Craver and Bechtel's (2007, p. 552) introduction of the maxim that causes temporally precede their effects:

> Many theories of causation assume that causes precede their effects. This feature of causation is often disputed (see Faye 2005), and some accounts of causation (e.g., Reichenbach 1958) are designed as the foundation of an account of the temporal order, and so do not assume the temporal asymmetry of causation.

The Faye source cited here (Faye, 2005) is to the then latest, 2005 Stanford Encyclopedia article "Backward Causation." (The article has since been revised in several later editions.) It constitutes a rather substantial dispute of the maxim. The article responds to the suggestion that backward causation is somehow paradoxical and reports several physical theories that propose backward causation.[16]

In another context, Frisch (2009a, §2) has argued for a similar principle in the context of dispersion processes in classical electrodynamics: "an effect cannot temporally precede its causes." A difficulty with the proposal is noted in Norton (2009). It is that the physics of dispersion in classical electrodynamics is time reversible. That means that, if we have a process in which state or event A precedes B in time, then we can also have the time-reversed process in which B precedes A. If we judge that A causes B in the first process, then we must also judge that A causes B in the second time-reversed process. But that relation in the time-reversed case would be causation backward in time. An unsuccessful escape from this conclusion is the empirically unfounded assumption that there are time-directed causal facts in classical electrodynamics that elude its laws.

After affirming the maxim that causes must precede their effects, Craver and Bechtel (2007, p. 553) turn to the second maxim of the asymmetry of causation:

> [The maxim that causes must precede their effects] raises a related worry about the asymmetry of causation.[17] It is a widely accepted condition on accounts of causation that they account for the asymmetry of causal dependency. The sun's

---

[16] The Reichenbach citation is harder to characterize as a dispute. Reichenbach (1956, §21) introduces the "causal theory of time," in which one event is *defined* as later in time than a second event, if that second event can causally influence the first event. A unidirectional sense of time arises in this theory only if the causation relation is asymmetric.

[17] Craver and Bechtel's footnote: "Again, this principle has been questioned. See Price (1996) for a lengthy review."

elevation causes the length of the shadow, but the length of the shadow does not determine the elevation of the sun. The virus produces the spots on the skin, but the spots on the skin do not cause the infection with the virus. Causes produce their effects, and (at least in many cases) not vice versa. Examples such as these have the staying power that they do because the asymmetry of causation is so fundamental to our very idea of causation.

The work of Price (1996) referred to in the footnote to the text does more than merely question the asymmetry of cause and effect. It proposes that it is an artifact of our human conceptions and not a factual matter in the world (p. 161): "On the perspectival view, causal asymmetry reflects an asymmetry in us, not an asymmetry in the external world." Price (1996, pp. 183–7) also uses the time reversibility of elastic collisions of billiard balls in the same way as I did above with the time reversibility of classical electrodynamic processes. Since the collisions are time reversible, there is nothing in the physics itself that distinguishes which state is the cause and which the effect.

More generally, the asymmetry of cause and effect appears sustainable only as long as we choose favorable examples. It is hard to retain this asymmetry when one considers equilibrium chemical reactions. For example, hydrogen and iodine combine to form hydrogen iodide:

$$H_2 + I_2 \rightarrow 2HI$$

Conversely, hydrogen iodide can decompose into hydrogen and iodine:

$$2HI \rightarrow H_2 + I_2$$

We may fill our reaction vessel with hydrogen and iodine and the first reaction will start to produce hydrogen iodide. If we hold the temperature at 445°C, the production will cease at 79% hydrogen iodide and 21% elemental hydrogen and iodine. If instead we had initially filled our reaction vessel with hydrogen iodide, the second reaction will start to form hydrogen and iodine. In neither case will the reactions go to completion. Rather, holding the temperature at 445°C, they will approach a dynamic equilibrium at the percentages above in which both reactions proceed but the rates of formation of reaction products of one reaction will match the rates of consumption in the other.[18]

One can describe this dynamical equilibrium as the occurrence of two asymmetric causal processes at the same time. In one, the cause, hydrogen and iodine,

[18] These reactions and conditions are as reported in Senter (1913, p. 164). Senter notes also that "The investigation of this reaction has been of considerable importance in the development of chemistry..."

leads to the effect, hydrogen iodide, through the first reaction. In the other, the cause, hydrogen iodide, leads to the effect, hydrogen and iodine in the second reaction. That is an artificial division. In this dynamical equilibrium, the reagents hydrogen, iodine, and hydrogen iodide are both causes and effects at the same time. The dynamical equilibrium is the continuing execution of the two reactions above. However, if any causal account is appropriate, it would be one in which we say that these reagents are engaging in a continuing causal interaction, without the asymmetry of the distinction of cause and effect playing any role. All the reagents are at once both cause and effect.

Craver and Bechtel are concerned that allowing causal processes to proceed in both directions would lead to troublesome causal circles (2007, p. 552): "the possibility of bottom-up and top-down influence 'propagated' simultaneously across levels results in problematic causal circles." Perhaps this may happen in specific cases. However, as a general matter, there can be no maxim that prohibits causal circles. For the chemical equilibrium described above is just such a causal circle. Equilibrium systems with analogous causal circles are pervasive in chemistry and in other sciences. Woodward (2021, pp. 15–16) finds nothing incoherent in such causal cycles and reports their existence in biological, social, and economic systems. They are essential, he notes, in control systems that employ feedback loops.

10. Conclusion

This chapter has presented a sustained critique not of causation, but of the metaphysical literature on causation. Its basis is empirical. Things connect with other things in the world in a myriad of interesting ways. It is the job of empirical science to find out just how they connect. Empirical science has enjoyed immense success in these efforts; and we can expect its successes to continue without any foreseeable limit.

As these successes continue, one feature of them persists: we cannot know just what the next discoveries will be. Non-empirical attempts to predict them have not and will not end well. Yet just such attempts have been the enduring substance of the metaphysics of causation. Its history is one of the repeated failure of efforts to impose enduring factual restrictions. By demanding that future science must respect the present conceptions of causation, they are attempts to circumscribe what empirical science can and cannot discover next. They have been, in effect, ill-fated attempts at a priori science.

The repeated history of errors and corrections in the metaphysics of causation is quite unlike the history of corrections in science. The power of discovery and correction derives from the empirical probing of the world that lies at the center of science. It has within itself the means to correct its own errors.

Metaphysical investigations into causation have no corresponding power. They can only pronounce and then accept their errors when subsequent discoveries in science reveal them.

If the analysis of causation cannot add anything factual to what science tells us, what can it provide? I have argued here that the identification of some things as causes and effects and some processes as causal is a matter of conventional definition. While this conception has lowered ambitions, it can be valuable if the designations are useful to us. Time and again, conceiving a process in causal terms calls up analogies that aid our limited human brains in comprehending novel processes. The interventionist approach provides special benefits. For if we know that a process is causal in the interventionist sense, we know that we can intervene in it and use it to manipulate things in the world. Since the account is purely one of definition, there is no presumption that Nature must present us with processes that admit interventions. But when they are found, the valuable discovery is promulgated efficiently by the simple expedient of attaching the label "causal" to them.

References

Audi, Robert, ed. (1999) *Cambridge Dictionary of Philosophy*. 2nd ed. Cambridge: Cambridge University Press.

Boyle, Robert (1688) *A Disquisition about the Final Causes of Natural Things*. London: John Taylor.

Craver, Carl F. and Bechtel, William (2007) "Top-Down Causation without Top-Down Causes," *Biology and Philosophy*, **22**, pp. 547–63.

Descartes, René (1982) *Principles of Philosophy*. Translated by V. R. Miller and R. P. Miller. Dordrecht: Kluwer.

Dowe, Phil (2000) *Physical Causation*. Cambridge: Cambridge University Press.

Earman, John (1986) *Primer on Determinism*. Dordrecht: Reidel.

Einstein, Albert (1917) "On the Quantum Theory of Radiation," pp. 63–77 in B. L. van der Waerden, ed., *Sources of Quantum Mechanics*. Amsterdam: North-Holland, 1967.

Ellis, George F. R. (2009) "Top-Down Causation and the Human Brain," chapter 4 in N. Murphy, G. F. R. Ellis, and T. O'Connor, eds., *Downward Causation and the Neurobiology of Free Will*. Berlin: Springer.

Faye, Jan (2005) "Backward Causation," in *The Stanford Encyclopedia of Philosophy* (Fall 2005 Edition), ed. Edward N. Zalta, https://plato.stanford.edu/archives/fall2005/entries/causation-backwards/. Stanford, CA: Stanford University.

Fisher, Ronald A. (1934) "Indeterminism and Natural Selection," *Philosophy of Science*, **1**, pp. 99–117.

Fleming, William (1860) *The Vocabulary of Philosophy*. 2nd ed. Philadelphia: Smith, English and Co.

Frisch, Mathias (2009a) "'The Most Sacred Tenet'? Causal Reasoning in Physics," *British Journal for the Philosophy of Science*, **60**, pp. 459–74.

Frisch, Mathias (2009b) "Causality and Dispersion: A Reply to John Norton," *British Journal for the Philosophy of Science*, **60**, pp. 487–95.

Gillett, Carl (2017) "Scientific Emergentism and Its Move beyond (Direct) Downward Causation," chapter 14 in M. P. Paoletti and F. Orilia, eds., *Philosophical and Scientific Perspectives on Downward Causation*. New York: Routledge.

Hitchcock, Christopher, "Probabilistic Causation," in *The Stanford Encyclopedia of Philosophy* (Spring 2021 Edition), ed. Edward N. Zalta, https://plato.stanford.edu/archives/spr2021/entries/causation-probabilistic/. Stanford, CA: Stanford University.

Hume, David (1739) *A Treatise of Human Nature*. Edited by L. A. Selby-Bigge. Oxford: Clarendon, 1896.

Kim, Jaegwon (1989) "Mechanism, Purpose, and Explanatory Exclusion," *Philosophical Perspectives*, **3**, pp. 77–108.

Kim, Jaegwon (1998) *Mind in a Physical World: An Essay on the Mind-Body Problem and Mental Causation*. Cambridge, MA: Bradford, MIT Press.

Kim, Jaegwon (2005) *Physicalism, or Something Near Enough*. Princeton: Princeton University Press.

Kistler, Max (2017) "Higher-Level, Downward and Specific Causation," chapter 4 in M. P. Paoletti and F. Orilia, eds., *Philosophical and Scientific Perspectives on Downward Causation*. New York: Routledge.

Laplace, Pierre-Simon (1805) *Traité de Mécanique Céleste*. Vol. 4. Paris.

Laplace, Pierre-Simon (1902) *A Philosophical Essay on Probabilities*. London: John Wiley & Sons.

Lowe, E. J. (2003) "Physical Causal Closure and the Invisibility of Mental Causation," chapter 6 in Sven Walter and Heinz-Dieter Heckmann, eds., *Physicalism and Mental Causation*. Exeter, UK: Imprint Academic.

Mackie, John L. (1980) *The Cement of the Universe: A Study of Causation*. Oxford: Clarendon Press.

Mellor, David Hugh (1995) *The Facts of Causation*. London: Routledge.

Menzies, Peter and Price, Huw (1993) "Causation as a Secondary Quality," *British Journal for the Philosophy of Science*, **44**, pp. 187–203.

Menzies, Peter and Beebee, Helen (2020) "Counterfactual Theories of Causation," in *The Stanford Encyclopedia of Philosophy* (Winter 2020 Edition), ed. Edward N. Zalta, https://plato.stanford.edu/archives/win2020/entries/causation-counterfactual/. Stanford, CA: Stanford University.

Mill, John Stuart (1882) *A System of Logic*. 8th ed. New York: Harper & Bros.

Newton, Isaac (1761) *Four Letters From Isaac Newton to Doctor Bentley*. Pall Mall: R. and J. Dodsley.

Norton, John D. (2003) "Causation as Folk Science," *Philosophers' Imprint*, **3** (4).

Norton, John D. (2009) "Is There an Independent Principle of Causality in Physics?" *British Journal for the Philosophy of Science*, **60**, pp. 475–86.

Norton, John D. (2010) "Cosmic Confusions: Not Supporting versus Supporting Not-," *Philosophy of Science*, 77, pp. 501–23.

Norton, John D. (2014) "A Material Dissolution of the Problem of Induction," *Synthese*, **191**, pp. 671–90.

Norton, John D. (2016) "Curie's Truism," *Philosophy of Science*, **83**, pp. 1014–26.

Norton, John D. (2021a) "Eternal Inflation: When Probabilities Fail," *Synthese*, **198** (Suppl. 16), S3853–75.

Norton, John D. (2021) *The Material Theory of Induction*. BSPSopen/University of Calgary Press.

Norton, John D. (forthcoming) *The Large-Scale Structure of Inductive Inference*. BSPSopen/University of Calgary Press.

Osler, Margaret J. (1996) "From Immanent Natures to Nature as Artifice: The Reinterpretation of Final Causes in Seventeenth-Century Natural Philosophy," *The Monist*, 79, pp. 388–407.

Paoletti, Michele Paolini and Orilia, Francesco, eds. (2017) *Philosophical and Scientific Perspectives on Downward Causation*. New York: Routledge.

Pearl, Judea (2009) *Causality: Models, Reasoning, and Inference*. 2nd ed. Cambridge: Cambridge University Press.

Price, Huw (1996) *Time's Arrow and Archimedes' Point*. Oxford: Oxford University Press.

Reichenbach, Hans (1958) *The Philosophy of Space and Time*. New York: Dover Publications.

Reichenbach, Hans (1956) *The Direction of Time*. Edited by Maria Reichenbach. Berkeley: University of California Press.

Russell, Bertrand (1912–13) "On the Notion of Cause," *Proceedings of the Aristotelian Society*, **13**, pp. 1–26.

Salmon, Wesley (1984) *Scientific Explanation and the Causal Structure of the World*. Princeton: Princeton University Press.

Senter, George (1913) *A Text-Book of Inorganic Chemistry*. 2nd ed. London: Methuen & Co.

Waismann, Friedrich (1959) *Turning Points in Physics*. Amsterdam: North-Holland.

Woodward, James F. (2003) *Making Things Happen: A Theory of Causal Explanation*. Oxford: Oxford University Press.

Woodward, James F. (2016) "Causation and Manipulability," in *The Stanford Encyclopedia of Philosophy* (Winter 2016 Edition), ed. Edward N. Zalta, https://

plato.stanford.edu/archives/win2016/entries/causation-mani/. Stanford, CA: Stanford University.

Woodward, James F. (2021) "Downward Causation Defended," pp. 217–51 in J. Voosholz and M. Gabriel, eds., *Top-Down Causation and Emergence*. New York: Springer.

Wysocki, Tomasz (manuscript) "Underdeterminsitic Causation: A Proof of Concept."

Zenneck, Jonathan (1901) "Gravitation," in *Encyklopädie der mathematischen Wissenschaften mit Einschluss ihrer Anwendungen*, Vol. 5, Part 1 (1903–21), pp. 25–67.

5

A Dispositional Account of Causation

Implications for the Biological Sciences

Rani Lill Anjum and Elena Rocca

1. Introduction

Does it matter whether causation is understood as regularities, difference-makers, mechanisms, or dispositions, or is this a purely philosophical debate that scientists and others can ignore? Some see philosophy and science as clearly separated, especially if one thinks of science as a largely empirical matter that might as well stay clear of metaphysics. There is a long tradition in philosophy of separating science from metaphysics, with Francis Bacon, the British empiricists, and the positivists as central historical figures. From this perspective, any speculations about the nature of causation should be left to philosophers, and empirical investigations of causation should be left to scientists. Our answer here is a different one, namely that one's philosophical understanding of causation has some very practical implications for the empirical sciences.

Scientific methods, we argue, come with some implicit philosophical assumptions about the nature of causation, but also of probability, complexity, and risk (Kerry et al. 2012, Rocca and Andersen 2017, Anjum and Mumford 2018a, Anjum and Rocca 2019, Andersen et al. 2019, Anjum et al. 2020). We will here show how philosophical thinking about causation affects the biological sciences in some very practical ways. Specifically, we look at the problem of external validity and how risk predictions sometimes fail when results are applied beyond the context of study. By adopting a dispositional account of causation, we argue that the methods for studying causal complexity and interactions in standard risk assessment ought to change. For this purpose, we also present our preferred version of a dispositional account of causation, named Causal Dispositionalism. The theory was first developed in detail by Mumford and Anjum (2011a) in *Getting Causes from Powers*, and most of its distinct features are found here. Starting from an ontology of dispositions, or causal powers, causation is on this account understood as something singular, complex, context-sensitive, and irreducibly tendential. Some of these features have been explained and developed in more detail in later works (e.g. in Anjum and Mumford 2017, 2018a, 2018b, 2018c).

Rani Lill Anjum and Elena Rocca, *A Dispositional Account of Causation: Implications for the Biological Sciences*
In: *Alternative Approaches to Causation: Beyond Difference-making and Mechanism*. Edited by: Yafeng Shan,
Oxford University Press. © Oxford University Press 2024. DOI: 10.1093/oso/9780192863485.003.0005

Causal Dispositionalism is proposed as a radical departure from the standard conception of causation, which we argue is largely premised on the two-event model, inspired by David Hume's analysis. Causal Dispositionalism is thus offered as an alternative to Hume's regularity theory, but also to anti-Humean alternatives that primarily see dispositions as the ontological basis for causal regularities.

2. The Two-Event Model of Causation as a Philosophical BIAS

Hume (1739, 1748) analysed causation as a relation between two events that are spatially and temporally conjoined, but otherwise unconnected. A 'Humean world' is thus a disunited world that consists in a succession of distinct, discrete events that appear in a more or less regular manner. This is known as the Humean mosaic (Lewis 1973/1986; see also Mumford 2004, 2005). Some of these regularities might be causal, but most are not. The epistemic challenge is then to decide which of the many regularities are in fact causal, which is a task that scientists spend much energy pursuing. This is the basis for the two-event model of causation, which can be detected in most philosophical accounts.

Mirroring Hume's famous billiard balls, a standard way to illustrate causation is as two nodes, A and B, with an arrow going from A to B (Figure 5.1). The model sits well with the Humean premise that causation is an unobservable relation (the arrow) between two observable events (the nodes A and B) that are spatio-temporally conjoined. One philosophical debate inspired by this model is which type of causal relation the arrow is supposed to represent. Various suggestions have been offered: perfect regularity (Psillos 2002), difference-making (Collins, Hall, and Paul 2004), counterfactual dependence (Lewis 1973/1986), physical necessity (Kutach 2007), probability-raising (Suppes 1970, Mellor 1995), mechanism (Glennan 1996, 2009), action (Gillies 2005), manipulability (Woodward 2003), physical transference (Dowe 2000, Kistler 2006), dispositions (Harré and Madden 1975), production, or some combination of these (Hall 2004, Cartwright 2004, 2007). Another philosophical debate is how to ensure, theoretically, the perfect regularity of A and B. One option is that the cause is taken as sufficient for the effect, which typically requires the addition of a *ceteris paribus* clause, indicating some normal or ideal conditions under which B would follow from A. Mackie's (1980) INUS conditions are a variant of the sufficiency view, in which A refers to a whole set of sufficient conditions. A second option is to see A as a necessary condition for B, so that without A, B wouldn't follow, which is the counterfactual dependence view (Lewis 1973).

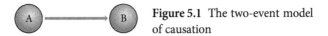

Figure 5.1 The two-event model of causation

The standard dispositional account follows the same lines, but usually translated into the stimulus–response model. The arrow is the causal relation between two observable events A and B, where A is some stimulus and B is the manifestation of a disposition. If the disposition is the flammability of a match, then the cause is the striking of the match and the effect is that the match lights. The disposition thus adds an ontological element that makes the causal claim true: 'striking a match causes it to light (under the right conditions)'. The disposition could then be used to explain perfect regularities, difference-making, necessity, mechanism, and so on, without rejecting the basic account. While the Humean regularity view is that no further connection is needed for causation beyond the perfect regularity of A and B (see Beebee 2000, 2006), anti-Humeans require some extra layer of necessity, such as a law or a disposition, for a regularity to count as causal (see Mumford 2005). Some dispositional accounts favour a 'conditional' form of necessity, in which the effect is necessitated when all the conditions are right (e.g. Marmodoro 2016, Heil 2017). The common philosophical idea that causation involves perfect regularity plus necessity is also part of what Norton (2007) calls the 'folk theory' of causation: 'the same causes *always* bring about the same effects' (see also Kutach 2007).

Returning to the opening question, whether it matters for science if the causal relationship is understood in terms of perfect regularities, difference-makers, laws, or dispositions, it seems the answer would be 'not much'. As long as the theory stays within the two-event model, the main disagreement is about matters that don't affect scientific methods or practice, or at best inform which language to use when reporting the causal findings. Regardless of whether one prefers to describe causation in terms of dispositions, mechanisms, laws, interventions, or perfect regularities, the methods and type of evidence required could stay the same. Any of these concepts might be compatible with difference-making, for instance. Striking the match makes a difference to it lighting, whether it is because of a disposition, a physical law, or a mere regularity. What indeed matters for science is how the two-event model dictates a certain type of approach that we might call 'the standard scientific methodology'. We will return to this methodology in biological research later in the chapter, but for now it should suffice to give a brief outline of its general features.

Following the two-event model, the standard scientific methodology is tailored to infer a causal relationship from the observations of A and B using more or less sophisticated tools. The method might require many repeated observations of A and B (regularity), or it might require a control to check whether B occurs also without A (counterfactual dependence). Causal complexity must then be reduced, to avoid overdetermination (difference-making), meaning that each individual factor must be separated and studied in isolation whenever possible in order to identify its causal role. Replications of the 'normal' or 'ideal' (*ceteris paribus*) conditions for the effect can be produced artificially, or theoretically, using models and controlled settings (idealization), which also makes reliable predictions possible. This presupposes that causal complexity is composed additively, from multiple elements that maintain their identity and causal role across contexts and independently of one another (Humean mosaic).

The artificial and isolated causal set-up that is typical for lab contexts and idealized models is what Cartwright (1999, p. 50) famously refers to as a nomological machine, designed to produce perfect regularities either in theory or in practice. The nomological machinery of science is very different from the natural contexts to which the results are supposed to apply, which are complex, messy, and local (Cartwright 2007, 2020). Along similar lines, Dupré (1993) talks of the 'disorder' of nature. He argues that any perfect, clear-cut classification of nature is artificial and mistakenly builds on a substance metaphysics of discrete things that have determinate boundaries and essences (Dupré 1981, 2007; see also Dupré and Nicholson 2018). The perfect regularities assumed in most accounts of causation might not ever happen in science, therefore, if it wasn't for the massive effort to keep the conditions for the effect artificially ideal, using theoretical models, controlled experimental settings, ideal abstractions, and causal isolation.

When scientific methods and practices are motivated by implicit philosophical assumptions, it is a form of bias. We call it a philosophical BIAS (= Basic Implicit Assumption in Science), but they are similar to other biases in that they tacitly skew 'the development of hypotheses, the design of experiments, the evaluation of evidence, and the interpretation of results in specific directions' (Andersen et al. 2019, p. 1). Philosophical biases are usually adopted through science education, and can be ontological, epistemological, or norms of methodological practice. Unlike other biases, philosophical biases are an integral part of science and cannot be avoided completely. One cannot, for instance, choose to make no assumption about the nature of causation, probability, or complexity in one's methodology. Instead, researchers can be taught to recognize, explicate, and critically reflect upon their own philosophical biases. As a result of such reflection, one can end up replacing one philosophical bias with another—for instance, if one replaces the assumption of dualism with reductionism or holism.

The two-event model of causation is a good candidate for a philosophical bias in science because it dictates what can count as causation and also which types of method to use. One example is the evidence hierarchy in medicine (Guyatt et al. 1992, Sackett and Rosenberg 1995, Sackett et al. 1996), which dictates that the best causal evidence is to establish that an intervention is a statistical difference-maker when comparing correlation data from a test population against a control population (randomized controlled trials). The second-best form of causal evidence is statistical correlations (cohort and case-control studies), and at the bottom is mechanistic evidence, which is theoretically motivated and generated from a variety of methods, including for example lab models (mechanistic studies). Translating this to a philosophical bias, we can say that the evidence-based hierarchy reveals the philosophical bias that causation is a regularity plus difference-making, or even manipulability (interventionist theory), which may or may not involve some underlying mechanism. In biology, however, correlations and difference-makers are considered weak evidence of causation if not supported by mechanistic evidence. The same, it could be argued, is needed for medicine, where mechanistic knowledge is often used to establish causal hypotheses, explain disease, develop cures, and evaluate the safety of treatment (Gillies 2018).

A causal mechanism can refer to a theoretical explanation or element of how the effect is produced (Illari 2011). An example is the way in which ibuprofen reduces inflammation by blocking the COX2 enzyme and synthesis of inflammatory prostaglandins. A mechanism thus refers to the causal pathway of how one gets from the cause to the effect, which then explains an observed regularity. Mechanisms are not included in Hume's analysis of causation. First, they go beyond the data and refer to theory, and second, they might be used to introduce an element of necessary connection. In each case, mechanistic explanations fail to meet the empiricist criterion. Even so, the standard mechanistic view of causation is perfectly compatible with the two-event model, since the arrow typically represents the mechanism linking the cause to the effect: $C \rightarrow M \rightarrow E$.

3. Implications for Biology: The Problem of Low External Validity

The standard scientific methodology, we saw, reveals a number of assumptions about the nature of causation, supported by the two-event model. We will now look at some practical implications that this approach has for biological research, specifically for external validity. When results are applied beyond its context of study, they sometimes fail to make accurate predictions. One reason for this is that, in order to generate repeatable results, the methodology requires artificial set-ups that are very different from the real-life contexts in which those results are needed. We will look at three examples from various areas of biological research: medicine, gene technology, and ecology.

3.1 The Translation Crisis in Medicine

In medicine, a problem of external validity is found in the so-called 'translation crisis'. This term refers to the fact that the large amount of resources invested in medical research does not result in an equally large improvement of people's health. In fact, a big part of medical knowledge that is gained with lab experimentation seems to be systematically lost when applied to patients. Many express this as a mismatch between the *benchside*, or pre-clinical knowledge, and the *bedside* and *community*, or clinical application (Cohrs et al. 2015).

To explain the problem of knowledge translation in medicine, we need to understand the meaning and purpose of the distinction between clinical research, which involves humans, and preclinical research, which involves lab models. The path between the design of a potential medicine and its marketing is long and strictly standardized. Efficacy and safety of medicines are evaluated in lab and computer models before they are approved for the clinical phase, meaning the phase in which medicines are tested for their use in human patients. Even then the clinical phase proceeds by gradual steps, in which the medicine is first tested for serious side-effects on healthy participants (clinical phase I), then tested on a small number of patients to establish the optimal testing doses, before finally being tested for safety and efficacy in large clinical trials.

For more than 15 years, there has been accumulating evidence showing that only a small percentage of the chemical compounds that are tested in preclinical studies ever receive marketing approval, and the extent of the phenomenon is so great that the preclinical to clinical translation has been called the 'valley of death' (Seyhan 2019). The reasons for this failure are, of course, complex and span from the more technical to the more socio-political level, including funding and incentive systems. Among these, a much-discussed issue that is of interest for us here is the clinical relevance of basic research and lab experimentation. A typical scenario, indeed, is that laboratory experiments test the propensity of a medicine to interact with a specific target under specific conditions, in order to interfere with the mechanism of the disease's onset. Often, however, only a subset of the patients has the target and the specific conditions required for the medicine to work. The illness in a real setting is generated by a complex intertwining of factors and conditions, and the task of modelling this in preclinical studies is a challenging one.

The traditional approach to this challenge has been to improve laboratory research by increasing precision and rigour. This means that one ought to make sure that the outputs of the experiment show, in a reliable way, that the medicine in fact does make a difference to each of the relevant factors. For this, one has to stick to the principle that when designing an experiment, one needs to control all variables except the treatment (Festing 2010). For this reason, pharmacological studies use standardized conditions and animal models that are selected to be genetically identical. The suggestion, in other words, is that basic science ought to stick to the best scientific practice, and that more precise results on the preclinical side will ultimately result in a better translation of knowledge to clinical research. However, an increase of precision, which must typically be obtained by causal isolation, comes as a trade-off with external validity. Results end up being reliable in the models and conditions of the test, but it becomes difficult to predict how reliable they are under other conditions. This can be seen when results are applied from one species to another, but also within the same species from one strain of animal models to another, and even with the same animal strain from one lab to another. Let us look at each of these cases in a bit more detail.

Application of Results from One Species to Another

Phylogenetic differences between different species make it difficult to predict impacts of medical treatments in humans from the way they work in animals. This is a well-known challenge, and also the reason why medicines are sometimes tested in more than one species in preclinical trials. Increasing the number of models, or standardized conditions, is however not a guarantee of success. An example is the medicine thalidomide, today used to treat leprosy and some forms of cancer, and infamously known for provoking major malformations in the developing embryo when first commercialized as an antidepressant in the late 1950s (Dally 1998). One problem is that the embryonic toxicity was overlooked because of the lack of strict safety regulation at the time. Indeed, the medicine was never administrated to pregnant animals before marketing, and nevertheless was used broadly by pregnant women with the off-label purpose of alleviating morning sickness. But a further problem, more relevant for us here, is that the embryonic malformations caused by thalidomide in humans cannot be easily detected in rodents, which were normally used as models for pharmacological tests. Different impacts of thalidomide can be detected in developing embryos of different species, such as rabbits and primates (Vargesson 2013).

Application of Results from One Strain to Another

In order to minimize complexity and control all the factors that could influence the result, pharmacologists use animal models that are highly genetically homogeneous. This means that, through inbreeding and selection, experimental animals belonging to the same strain are genetically very similar to one another. This way, one makes sure that any observed difference of effect between the medicine and the control is because of the medicine and not genetic variation. The problem with this is that experiments in one inbred animal strain sometimes have low predictive power with respect to the results in other inbred strains. Indeed, different strains have different susceptibility to different chemicals, or to different genetic mutations. For instance, the deletion of the same gene from the rat genome could provoke pre-natal death in one mouse strain, while leading to a normal span in another (Kacew et al. 1995).

Application of Results from One Lab to Another

More surprisingly, perhaps, the same experimental set-up can give different results depending on the lab environment, even when the same animal strain is used. Rather contra-intuitively, extreme standardization seems to worsen, rather than increase, the chances that the same results are repeated from one lab to another. In fact, it has been shown that, to secure better trans-lab repeatability, it is better to increase the variation, rather than to standardize the experimental protocol (Richter et al. 2011). This is not very surprising today, since we know that the environment influences the diversity of gut microbiota, which in turn influences protection and priming of immune systems. Thus, mice from different lab environments can have different impact from exposure to the same chemical, even if they are genetically identical (Jakobsson et al. 2015).

3.2 Stability and Adaptation in Gene Technology

Worldwide, agriculture makes a preponderant use of transgenic crops. These are crops in which certain gene fragments were introduced from other species (usually bacteria or other plants), in order to confer an advantageous trait. For instance, the so called '*Bt maize*' is a type of genetically modified corn that produces the insecticidal Bt protein, so that insects who try to feed on the corn die without the need of spraying insecticide on the plant. This property has been developed by inserting, in the corn genome, the gene for the Bt protein, which is originally obtained from the bacterium *Bacillus thuringiensis*.

When a new gene is artificially introduced in a genome, one should expect that the genome becomes unstable, and some genetic rearrangements will happen as a result. Because of this, every genetically modified (GM) plant must be assessed for molecular stability *before* it enters the market. This means that, before obtaining permission to commercialize the GM crop, the developer needs to grow their product in experimental fields, under a variety of standardized conditions (e.g. temperature and humidity), and show that the characteristics of the plants remain stable through generations. However, there is no requirement for this type of monitoring in commercial GM hybrids after they are sold to farmers and culti-vated in different environments. This means that the seeds are sold with a guarantee of a certain yield, based on the yield and the stability observed in the experimental field. However, assessing the real-world performance of the trans-genic modification is not a straightforward task, and there is controversy about whether the guarantee of a higher yield, used by the GM crop producer as a marketing strategy, corresponds to the reality (Hicks 2015). On one side, there is the argument that one needs experimental isolation to establish that a change in yield is due to the transgenic modification.

> To determine the contribution of these transgenes to yield, research must be able to isolate their effects from the many other factors that influence yield.... For studies to accurately attribute yield increases to transgenes, they must try to control or account for these factors.   (Gurian-Sherman 2009, p. 14)

Although practically impossible, such control of all factors except for the transgene is attempted in field trials. Here, the most obvious relevant variables are controlled for, such as the weather, agricultural practices, and pests. The question, however, is what these experiments tell us about the GM crops' performance outside experimentation. Here the opinions are split, mainly between the GM crop producers and the citizen groups and independent scientists. It is not our aim here to analyse this controversy in terms of contrasting social interests and basic assumptions, which was partially done elsewhere (Hicks 2015). Rather, we will describe why the reliability of GM manufacturers' predictions on future yield raises sceptical arguments.

When plants adapt to different environments, they change their physiological performance to maintain their fitness in the face of environmental stressors. The adaptation is a result of genetic and physiological changes, such as systemic acquired acclimation to strong light, which happens in some plants when they are exposed to potentially damaging full sunlight (Nancy and Eckardt 2007). With this system, the leaves that are exposed to sunlight can use the hormonal system to signal to shaded leaves and thereby initiate their pre-acclimatation. Clearly, when energy is used for adaptation systems like this one, less of it is available for yielding production. This reduces the external validity of pre-commercial experiments, based on controlled conditions. Indeed, the few available studies on the molecular stability of GM plants 'in the wild' seem to support that they have unintended traits, probably because of changes activated by the plant as a form of adaptation (Cellini et al. 2004). How exactly these rearrangements and physiological changes influence the performance of the transgenic modification remains so far unknown.

Application of results from field experiments to predict performance in other environments thus remains a problematic issue for gene technology. We have explained why this low external validity is at least partly due to standard scientific methodology that leaves a gap between the context of study and the context of application.

3.3 Predicting Ecological Impact

A chemical substance can be found safe in the lab, but the numerous ongoing lawsuits from local communities against governments and industry suggest that these conclusions cannot always be trusted (Wickson and Wynne 2012). This is because the exposure to a substance with some intrinsic toxicity, or *hazard* as it is called in technical language, will be more or less harmful depending on a number of contextual factors and conditions. Although it is uncontroversial that context plays a crucial role for toxicological risk, we argue that the implementation of context-based risk assessment frameworks is hindered by some methodological shortcomings (Rocca and Anjum 2019).

One example is toxicity from long-term exposure to residues from petroleum extraction. Petroleum is a complex substance, constituted by thousands of hydrocarbon compounds, some of these consisting of 100 carbon atoms or more. One issue is that two equally large patches of oil will probably have different composition and therefore different intrinsic toxicity or hazard. Another problem is that even residues with the same composition and intrinsic toxicity will tend to have different bioavailability (evaporate, dissolve, etc.), depending on environmental conditions. To face this complexity and create a tool which could allow us to predict the impact of petroleum exposure in living beings and ecosystems, the US Total Petroleum Hydrocarbon Criteria Working Group worked out a systematic, science-based guideline (Twerdok 1999). As in the previous examples, the standard scientific approach was once again based on decomposition of causal factors, analysis of these in isolation, followed by addition of the results. Let us look at how this was done in some more detail.

Petroleum was broken down into 13 groups, or 'fractions', where each fraction contained chemical components with the same size (or 'molecular weight', in chemical language). The assumption here is that compounds with similar molecular weight have similar physical-chemical properties. Some of the components in each fraction were then analysed based on their toxicity in animal models and on their chemical capacity of becoming bioavailable (Sawyer and Loja 2015). The idea behind this procedure is that even the most hazardous compound can cause no harm if living organisms remain unexposed to it. The chemical capacity of becoming unavailable, as well as the intrinsic toxicity of every component, were calculated using laboratory models and experiment. Typically, one would test for each component how much of a certain initial quantity can dissolve and evaporate under different standard conditions. These results were then modelled using mathematical tools to combine them with different kinds of geological conditions. To predict the bioavailability of a certain oil residue in a certain environment, the physical characteristics found in isolation for every component would be added together. Notice that this approach presupposes that chemicals combine in an additive way, excluding thereby the possibility of synergistic effects. This remains to date a common assumption in toxicology, and it was adopted for instance in the recent framework of RISK21, which applies it to chemicals with a common mechanism of toxicity and low levels of exposure, compatible with real-world situations. The aim of assessment is to identify a 'threshold of toxicological concern' for each specific mixture (Moretto et al. 2017). However, by overlooking potential synergistic effects there is a danger of underestimating the total toxicity of low exposure to multiple chemicals.

Such underestimation is precisely what seems to have happened in the case of the prediction of toxicity from petroleum. The result of the work from the US Total Petroleum Hydrocarbon Criteria Working Group was the identification of the lightest oil compounds that evaporated more easily than the ones with the highest acute toxicity. Despite their high intrinsic toxicity, heavier compounds were considered inert, not biologically available, and therefore safe. Consequently, oil toxicity was believed to be mainly acute, and to diminish exponentially with weathering. Later on, however, study designs with an ecological approach, based on the direct observation of ecosystems that were contaminated by petroleum (e.g. after oil spills), showed that the weathering of oil happens at an unpredictable pace that depends on the specific environmental conditions. Oil sediments can occasionally become physically trapped and therefore last longer. Compounds that were considered not bioavailable were found to be toxic even at very low doses in developing organisms. In the long term, oil effects in the population seemed to be amplified rather than diminished through environmental interactions, for instance in the trophic chain (Peterson et al. 2003).

Predicting ecological impacts of chemical mixtures from experimental isolation of each factor remains a challenge. As we will see later in the chapter, the shortcomings of the standard methodology have motivated a call for efforts to develop risk assessment frameworks that acknowledge the cumulative, non-linear composition of chemicals' toxicity (Callahan and Sexton 2007).

4. Diagnosis of the Problem: A Humean Methodology

As described above, the problem of low external validity for biological research typically happens when the context of study is very different from the context of application. Our diagnosis is that the standard scientific methodology is based on some Humean assumptions about causation that can only be fulfilled through theoretical abstraction or physical idealization. The lab context is a physical form of idealization that is used to artificially produce perfect regularities under some isolated and controlled conditions. This methodology is based on four assumptions about causation:

(i) the causal contribution of a single factor can be studied by isolating it from other factors (isolation);

(ii) the isolated cause should make an observable difference to the effect (difference-making);

(iii) the causal role is maintained across contexts (causal essence); and

(iv) the causal contribution observed in a semi-closed system can be used to make predictions about how it will behave in complex interactions in open systems (same cause, same effect).

One might argue that this methodology is a misinterpretation of the philosophical concept of causation, and that the change of context from closed to open system fails to preserve the *ceteris paribus* conditions: 'all else being equal'. The two-event model follows Hume in stipulating that the same cause gives the same effect *under the same conditions* (Hume 1739, p. 173). Once the conditions are different, however, one shouldn't expect that the effect follows. In fact, Hume said it even more strongly, that if there is an irregularity in the effect, then this must come from a difference in the cause (Hume 1739, p. 174). Our response to this is that, if causation is supposed to be perfect regularity plus something more, then scientists have no option but to try and produce perfect regularities in order to establish causation. In theoretical science, one might just stick to the idealized models for causal predictions as well. When those models fail to make accurate predictions, one can simply say that the results would have been otherwise had the conditions been ideal, or at least relevantly similar. In biology, however, this is rarely an option, especially for research that involves establishing and predicting the safety of medicine, genetically modified food, or chemical substances. If an intervention turns out to give harmful outcomes in real-life settings, it doesn't help much to say that it would have been safe if the conditions had been ideal. Simply acknowledging that all organisms have genetic variations should be enough to dismiss the ideal of science as a search for 'same cause, same effect' for any real-life contexts.

The low explanatory power of highly idealized models for real-life contexts is something that Nancy Cartwright has addressed repeatedly throughout her career. In a book that has become a classic within philosophy of science, *How the Laws of Physics Lie*, she argues that fundamental laws in physics fail to explain reality, and she emphasizes instead the need to ground scientific theory in local and concrete physical processes (Cartwright 1983). She suggests that laws and general causal claims are best understood as describing capacities, or causal powers (Cartwright 1994; see also Mumford 2004). Moreover, rather than deriving singular causation from generalized causal facts, Cartwright (1994) argues that we should take the single case as primary. We agree with all these suggestions. In what follows, we explain why we think Causal Dispositionalism is the best alternative for a dispositional account of causation if we want an alternative to the standard Humean conception.

5. Causal Dispositionalism

Causal Dispositionalism is an anti-Humean theory that emphasizes the causal role of properties (Mumford and Anjum 2011a). It starts from the ontological framework of dispositionalism, in which properties are understood as causally powerful. Dispositionalism is a neo-Aristotelian ontology, which has seen a revival in philosophy in recent decades (see, e.g., Harré and Madden 1975, Cartwright 1994, Mumford 1998, Mellor 1995, Ellis 2001, Molnar 2003, Heil 2004, Martin 2008, Marmodoro 2010, Bird, Ellis, and Sankey 2012, Greco and Groff 2013, Vetter 2015, Dumsday 2019, Jacobs 2017, Ingthorsson 2021). We here present

Anjum and Mumford's dispositional theory of causation. It cannot be a very detailed account, but we will outline some of the features that separate Causal Dispositionalism from theories of causation that are more in line with the two-event model.

The two-event model suggests that causation is an unobservable relation between two observable occurrences. In Causal Dispositionalism, however, causes are understood as intrinsic dispositions, or powers. Dispositions are properties that can exist unmanifested, which is why Hume dismissed them on empiricist grounds. There are, though, good reasons to see dispositions as real, even when they do not manifest themselves into some observable effect. As we said when discussing the case of petroleum toxicity, a chemical substance does not suddenly become toxic when it is bioavailable, or when it manifests its toxicity. To say that a substance is toxic must from a dispositionalist perspective refer to some intrinsic property of the substance rather than a mere correlation between exposure and observed outcome. Similarly, a dispositionalist will accept that someone can be fertile even if they live their whole life without bearing offspring, and that a powerplant can be explosive even if it never explodes. From a dispositionalist starting point, causation is what happens when dispositions manifest themselves (Mumford and Anjum 2011a, ch. 1).

If one wanted to make the dispositional account compatible with the two-event model, one could translate it into a stimulus–response model, where the stimulus and manifestation are the observable relata and the disposition is the unobservable truth-maker of the causal relation. In Causal Dispositionalism, however, this move is resisted, since causation is seen as an ontologically fundamental and primitive matter (Mumford and Anjum 2013, ch. 8). That causation is primitive means that it cannot be analysed into something else, such as constant conjunction, counter-factual dependence, mechanism, or even dispositions. A dispositional analysis of causation would be circular, since dispositions, or powers, are already causal (Mumford and Anjum 2011a, sect. 1.2). Instead of thinking of causation as a relation between two separated events, it is understood as a single temporally extended event, described as a continuous, unfolding process of change (Anjum and Mumford 2018b, pp. 67–8). Causation, on this theory, is a dynamic matter, typically involving change. The causal process begins when various dispositions come together and start interacting, and it takes time to unfold. This process can be very fast or very slow. For instance, if someone is exposed to a toxic substance, it might take only seconds before a harm is manifested, such as in the case of headache and drowsiness after exposure to light, volatile petroleum compounds. But it can also take years, as in the case of lung cancer from long-term exposure to cigarette smoke.

According to the two-event model, the default causal situation is that the effect is produced by a single cause, or the cause plus some right background conditions. In Causal Dispositionalism, however, the default expectation is that a causal process will involve multiple dispositions, or what is called mutual manifestation partners. These dispositions are not treated as background conditions, but as causes that must interact in order to produce the effect. We know, of course, that no medical treatment can do its causal work in total isolation, if left in the box. Instead, it requires causal interaction with an appropriate partner. For this, we need to consider not only the dispositions of the medicine, but also the dispositions of the person receiving it. If that person lacks the right biological receptor, they will not benefit from the treatment. The same holds for adverse effects. Someone might be a mutual manifestation partner for the targeted effect of a medicine, but not for any of its adverse effects. But the opposite might also happen, that one is not a mutual manifestation for the targeted effect of a treatment, but for one of the adverse effects. Thalidomide might have failed to alleviate morning sickness in some of the pregnant women while nevertheless contributing to foetal malformation, and vice versa. Note also that this harmful effect was not observed when the drug was tested on a number of animal models, as explained above.

It matters, therefore, which dispositions are changed once the context changes. According to Causal Dispositionalism, all causal processes are context-sensitive, and the degree of sensitivity depends on whether the dispositional manifestation partners interact in linear or non-linear ways. An example of a non-linear interaction is the combination of asbestos exposure and cigarette smoking, which drastically increases the risk of lung cancer compared to the risk from exposure to each of the substances alone. That a disposition will tend to produce different effects in different contexts means that one should not expect that the same cause will produce the same effect independently of what else is there. The *ceteris paribus* clause thus seems an indispensable, yet invisible, element of the two-event model.

Both the stimulus–response model and the two-event model fail to properly acknowledge the causal contributions of all the manifestation partners. Within the standard conception, causal complexity is dealt with in terms of background conditions. In Causal Dispositionalism, however, all dispositions in the background conditions are also causes, and any separation between conditions and causes is seen as purely epistemic, or perspectival (Mumford and Anjum 2011a, sect. 2.5). Causal complexity should not be understood as mere composition, or as an intricate web of extrinsic instances from which one could disentangle each causal factor and study it in isolation. Instead, new properties can emerge as a result of the complex interaction, which is what Anjum and Mumford (2017) call 'the causal-transformational model of emergence'. This means that the dispositions that the component factors have individually might disappear and transform into other dispositions during the process of interaction. As we saw above in the case of genetically modified crops, a genetically modified plant that adapts to new environmental stressors might acquire new traits, needed to maintain fitness despite of the stressors. At the same time, it might reduce or even lose the production of the genetically modified trait.

> With the causal-transformative account, emergent powers will not be a mere subset of the base powers. Until the causal transformation, those emergent powers do not exist in the base phenomena at all. So emergence involves a different set of powers; and once the causal transformation has occurred, the emergent powers exist only in the whole, rather than in the parts...
>
> (Anjum and Mumford 2017, p. 104)

The two-event model can be interpreted as a singular or general theory, but on Hume's account, causation is general. This means that causation requires repetition, or regularity, in order to count as such. In contrast, Causal Dispositionalism is a singularist theory. Causation happens in the concrete, single case because of the properties involved, and not because of some general covering law (Anjum and Mumford 2018a, p. 77). On this account, causation requires no regularity or repetition. This means that the account allows that in principle there could be a unique causal event where a type of disposition only manifests itself once. Since all living organisms are unique, for instance in virtue of genetic variations and exposure to different environments, one should not expect that the exact same mutual manifestation partner is ever entirely repeated, except in an artificially closed and controlled setting. So if someone gets exposed to a new medical treatment and dies, but no one else did, this should not in principle rule out that their death could be caused by that treatment. Instead, it becomes a question of identifying the causal mechanisms that could explain how some disposition of the treatment could cause such a biological reaction. It might have been the combination of the dispositions of the treatment and the dispositions of the individual, for instance if they had a rare allergy to one of the components. But it could also be the result of the interaction of two or more treatments that shouldn't be combined.

While repetition is usually part of an epistemic notion of causation, Causal Dispositionalism sees causation as an ontologically intrinsic, singular, and qualitative matter, rather than as a regularity plus (or minus) an extrinsic relation. Often, however, causation is assumed to be a perfect regularity precisely because it involves an extra element of necessity. According to Causal Dispositionalism, it is perfectly possible to have the cause without the effect (the full argument can be found in Mumford and Anjum 2011a, ch. 3). Causal production is thus not the same as causal necessitation (see also Anscombe 1971). The two-event model, however, suggests that whenever A happens, B follows. Such perfect regularity could be secured by simply stipulating that whenever the effect was produced, it was also necessitated. Necessity will then be guaranteed by some non-specific *ceteris paribus* clause, in which the conditions are sufficient for the effect, or sufficient 'in the circumstances'. Causal Dispositionalism dismisses this move, based on the idea that dispositions bring with them a form of modality that is short of necessity, yet more than pure contingency (Mumford and Anjum 2011a, ch. 8). On this view, causation is *irreducibly tendential.*

The dispositional modality involves directedness, intensity, and the possibility of counteraction (Anjum and Mumford 2018c, ch. 1). For instance, fertility is the tendency to bear offspring (direction), and someone can be more or less fertile (intensity). But no matter how strong the tendency is, all causal processes can in principle be counteracted by additive interferers (counteraction). One can remove something that takes away or reduces the disposition (subtractive interference) or add something that prevents it from manifesting (additive interference). The causal tendencies are not the same as statistical incidents or probabilities (Anjum and Mumford 2018a, sect. 9.4). Instead, they give rise to single propensities (Mellor 1971, Popper 1990). One might even interpret the need for a non-specified *ceteris paribus* clause as an admission of the deeply tendential nature of causation.

> The basics of this theory thus equip us to explain how predictions can often be reliable to an extent even though they are also fallible. Knowing that a cause tends towards a particular effect does not offer a guarantee; but it says more than that the effect is a mere possibility. The dispositional account also explains how the *ceteris paribus* clause that is often attached to predictions can be non-vacuously true. To say that shaken gelignite explodes, *ceteris paribus*, means that it has a real disposition to do so, even though there are circumstances in which it might not. (Anjum and Mumford 2018a, p. 189)

Replacing the two-event model of causation, Causal Dispositionalism favours a version of the vector model (Figure 5.2). The model allows the illustration of various features of causation (Mumford and Anjum 2011a, chs 2 and 4): the current situation (vertical line), including various causal powers (arrows) that have a certain direction and intensity (length); the resultant combined disposition (R), which can be a result of linear or non-linear interaction; and a threshold effect for some observable manifestation of interest (T). In addition, the model can be used to illustrate context-sensitivity, and the counteraction of the effect by additive or subtractive interference (Mumford and Anjum 2011b).

The vector model is itself an artificially closed and idealized set-up, of course, which might give the false impression that the resultant vector is necessitated by the specific combination of individual dispositions. In any real-life context, there will be too many dispositions involved to possibly include in a vector diagram, and there will be no complete understanding of how they interact. What the model can do, however, is to present a more realistic idea of how one cannot focus only on single causes (e.g. the stimulus or intervention), or only on those causes that

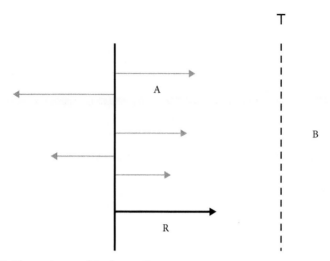

Figure 5.2 The vector model of causation

Table 5.1 Humean conception versus Causal Dispositionalism

Standard Humean conception	Causal Dispositionalism
Two-place relation	A single process
Static	Dynamic
Empiricist	Ontological
Events	Dispositions
Perfect regularity	Tendency
Background conditions	Causal mechanism
Single causes	Complexity
Causal laws	Singularism
Composition	Emergence
Difference-making	Mutual manifestation
Ceteris paribus clause	Context-sensitivity

dispose, or tend, towards the expected effect. In addition, one must consider how any new context will bring with it a unique set of mutual manifestation partners that must also be considered. Furthermore, the vector model shows how the resultant vector will depend not only on the combination of dispositions, but also on how they interact. Finally, it illustrates how one should expect that at least some of the dispositions involved could or in fact do counteract the predicted outcome.

Table 5.1 is an attempt to sum up some of the features of causation in Causal Dispositionalism and contrast them with the standard Humean conception, before we look at the methodological implications in the next section.

Returning to the problem of external validity addressed in section 3, for the biological sciences, we saw that results generated from artificially controlled lab-contexts often don't apply to the real-life contexts where the knowledge is needed. Given what we have said here about causal processes being highly sensitive to changes in the context, this should not surprise us. According to Causal Dispositionalism, the change of context involves a whole new set of mutual manifestation partners, and thus new causal interactions. We should therefore expect that different sets of dispositions will tend to produce different effects. The question is what can be done instead, to increase the external validity of biological research.

6. Methodological Implications

From a Causal Dispositionalist starting-point, scientific methodology should be quite different from the Humean methodology, indicated in section 4. In Table 5.2, we outline some of the ways in which these concepts of causation motivate different scientific approaches and focus. Among other things, Causal Dispositionalism suggests that we should steer our focus away from static, changeless factors, to be studied in separation and isolation within some artificially closed system, assuming additive composition of causes. Instead, the default expectation should be that causation is a complex and dynamic temporal process,

Table 5.2 Methodological implications of a Humean conception versus Causal Dispositionalism

Standard Humean conception	Standard methodology	Dispositionalist methodology	Causal Dispositionalism
Two-place relation	Correlation	Change	A single process
Static	Data	Processes	Dynamic
Empiricist	Observation	Theory	Ontological
Events	Causal relata	Potentiality	Dispositions
Perfect regularity	Statistics	Propensities	Tendency
Background conditions	Abstraction	Contextual	Causal mechanism
Single causes	Separation, isolation	Interaction	Complexity
Causal laws	Repetition, replication	Uniqueness	Singularism
Composition	Principle of additivity	Cumulative	Emergence
Difference-making	Controlled studies	*In situ*, local knowledge	Mutual manifestation
Ceteris paribus clause	Idealization	Variation, interference	Context-sensitivity

involving non-linear interactions within open systems. This also has implications for how we think of causal mechanisms as much more complex and sensitive to contextual influences than what can be detected when observed in isolation from any real-life contexts. If mechanisms are thought of as dispositional and tendential, then the more one knows about how something can be influenced by various types of dispositions, the better one can predict how something will behave in a new context, when interacting with different manifestation partners. Instead of placing the mechanism between the cause and the effect following the two-event model (C → M → E), we prefer Donald Gillies' (2018, p. 78) suggestion to reverse the mechanistic analysis of causation, to avoid circularity, and instead to account for mechanisms in terms of causation.

Like Gillies, we see causal mechanisms as involving complexity that can always be described with more detail added. Add to this that any real-life context is an open system, with potentially countless causal influences and interactions. This also means that mechanistic knowledge is never really complete (Gillies 2018, p. 181). In the same way, causal knowledge in the dispositionalist sense is also never complete, or possible to establish by using a single method. Instead, causal evidence requires a combination of methods, each of which will have certain strengths and limitations for establishing intrinsic dispositions. Elsewhere (in Rocca and Anjum 2020), we propose a dispositionalist pluralist approach to causal evidencing for medicine and public health, in which the various methods complement each other and fill in each other's blind-spots.

Causal Dispositionalism emphasizes the importance of contextual and local knowledge. To focus on single and isolated factors when making predictions about the outcome will be to disregard most of the causally relevant contributions in the causal interaction. Most predictions excluding complexity will therefore only apply to the model itself rather than to the contexts for which those predictions are needed. Note that causal complexity, uniqueness, and context-sensitivity are all features that one usually attempts to minimize for scientific purposes. Given the two-event model, this is no surprise, as we have already explained. Instead of seeing complexity, uniqueness, and contextual interference as obstacles for establishing causation, Causal Dispositionalism allows scientists to use these features to their advantage, as part of their methodology. If these are essential parts of causation, then we should expect that any natural causal process will include a unique set of multiple manifestation partners, some of which work with each other towards the effect and some that work against it. Rather than pretending that this is not the case, and simply stipulating some general but irrelevant causal claim, one can try to identify some of the unique dispositions that a certain local context or receptor brings to the causal situation, using for instance qualitative approaches and *in situ* studies. While it would be convenient for all real-life contexts to be identical to each other and to the model context, this is far from the reality. This is why scientists cannot be content with establishing causation and predicting risk only within an abstract model, relying on the conditions to be normal, average, equal, or ideal.

When assessing risk from a certain stressor, or exposure, one should seek to learn more about the full range of its causal powers. This cannot be done if it is only observed in isolation, since most of the possible manifestation partners with which it could interact would then be excluded from the study. The more we can observe the stressor in various contexts and combinations, the more we will learn about its dispositions. This, however, should be done within realistic contextual settings, where we can observe real causal processes in their full complexity under various conditions, and not in abstract or replicated idealized models where all possible interferers are removed. In practice, this means that we should study contexts of harm in detail, as a unique and valuable source of causal knowledge. Uniqueness is here a strength, since it can contribute to expanding our theoretical knowledge of the causal mechanisms, for instance if there is an unexpected outcome of a stressor when interacting with that unique manifestation partner. Detailed case studies of hazardous events might reveal unique or rare dispositions that left the situation more vulnerable to harm, which could then help us identify and possibly prevent future unwanted events. Whether it is a certain ecosystem that is particularly vulnerable, or an individual or population, the potential of new causal knowledge from one-off cases of harm remains the same. Drug safety, or pharmacovigilance, is one example where a detailed study of rare side-effects has the potential to motivate new scientific hypotheses and contribute to developing biological knowledge, which we call 'causal insights from failure' (Rocca et al. 2020).

Even though risk assessment approaches in pharmacovigilance, gene technology, and environmental risk have traditionally followed the standard Humean conception of causation, researchers from various disciplines have urged that a new methodology is adopted. In order to move beyond the assessment of single stressors, outcomes, sources, and pathways, they argue, one must acknowledge the crucial role of interactions between stressors and environmental factors for the insurgence of harmful effects in complex systems (Sexton 2015). For this, one needs to study synergism and inhibition for unidentified components as well, which is most accurately done by assessing the toxicity of whole mixtures (Boobis et al. 2011, Callahan and Sexton 2007). One also needs to focus on the receptor of exposure, which is the local context of application for the assessment results. Cumulative Risk Assessment was proposed as a framework for evaluating the combined harms from co-exposure to multiple stressors. It is not restricted to chemical stressors, but also includes biological, physical, and psychosocial entities. The purpose was to make risk assessment 'more reliable, realistic and relevant' (US EPA 2003, Williams et al. 2012). In Cumulative Risk Assessment, one should involve expertise and perspectives from social science and the local community, and use stakeholder involvement to gather better knowledge of the receptor (Callahan and Sexton 2007, Checker 2007, Dendena and Corsi 2015).

There is, however, a tension between feasibility and realism, as long as the concept of causation itself does not change. We have proposed three persisting challenges for a more dispositions-friendly risk methodology (Anjum and Rocca 2019):

(1) how to study causal complexity and interactions through separation, isolation, and addition;
(2) how to study all relevant causal knowledge through quantitative approaches and measures; and
(3) how to predict single propensities using statistics and probability theory.

Unless we change the Humean influences of scientific methodology, we argue, risk assessment methodology cannot evolve in the direction suggested. Indeed, more recent versions of Cumulative Risk Assessment, such as RISK21, seem to have returned to the standard scientific approach. Although one acknowledges that most risk factors interact in cumulative ways, the principle of additivity must be taken as the default assumption for pragmatic reasons. Factors linked to host, lifestyle, and environment are known to modulate the toxic effect of chemical stressors, but they add too much complexity to establish difference-making. Therefore, modulators should only be included 'if deemed necessary...and the number of variables to consider should be kept to a minimum' (Solomon et al. 2016).

When scientists get stuck in a methodology that they know doesn't work in practice, then something needs to change. We propose that what needs to change is the conception of causation and, with it, the standard scientific methodology that it motivates. Earlier, we said that scientific practice and methodology are motivated by basic implicit assumptions of a philosophical nature (philosophical BIAS), but now we see that these assumptions also hinder the development of better approaches. Even if one is not convinced by the dispositional account of causation, we hope to have shown that there is a version of it that would motivate a different scientific methodology. The standard Humean conception was based on strict empiricist criteria, whereby causal knowledge and evidence are restricted by what can be observed in actual occurrences. When assessing and predicting risk, however, we cannot wait for harm to happen before acknowledging something as potentially harmful. Theories of mechanisms, although they contain a speculative element, must play a central role in the assessment. Causal understanding, of *how* and *why* something might happen, is crucial for making safe decisions with a precautionary mindset, especially when trying to predict risk for open systems where causal complexity and the unknown unknowns are countless.

While the strict empiricist might say that what we don't know won't hurt us (at least as far as we know), the dispositionalist should expect that there are plenty of things that might hurt us, and might be hurting us at this moment, even if we haven't noticed it yet. The causal process might just take time to reach an observable threshold effect, at which stage it might be too late to counteract it, or it might just lack one or two manifestation partners before the full effect becomes evident to us. We suggest, then, that risk itself requires a dispositional rather than an empiricist interpretation, where potentials and underlying mechanisms are at least as important to us as observable occurrences when making predictions about harm. We also suggest that we need a more realistic risk assessment methodology, where one also studies real causal processes in their natural complex settings. Cumulative Risk Assessment is a methodology that scientists have promoted but failed to implement within the current approach to causation. It matters to science, therefore, what our philosophical concept of causation is. It matters even more to those affected by science if the methods fail to make accurate predictions about risk and safety for real-life settings.

References

Andersen, F., Anjum, R. L., and Rocca, E. (2019) 'Philosophy of biology: Philosophical BIAS is the one bias that science cannot avoid', *eLife*, 8: e44929, 10.7554/eLife.44929.

Anjum, R. L., and Mumford, S. (2017) 'Emergence and demergence', in M. Paoletti and F. Orilia (eds), *Philosophical and Scientific Perspectives on Downward Causation*, London: Routledge, pp. 92–109.

Anjum, R. L., and Mumford, S. (2018a) *Causation in Science and the Methods of Scientific Discovery*, Oxford: Oxford University Press.

Anjum, R. L., and Mumford, S. (2018b) 'A process theory of causation', in D. J. Nicholson and J. Dupré (eds), *Everything Flows: Towards a Processual Philosophy of Biology*, Oxford: Oxford University Press, pp. 61–75.

Anjum, R. L., and Mumford, S. (2018c) *What Tends to Be*, London: Routledge.

Anjum, R. L., and Rocca, E. (2019) 'From ideal to real risk: Philosophy of causation meets risk analysis', *Risk Analysis*, 39: 729–40.

Anjum, R. L., Copeland, S., and Rocca, E. (eds) (2020) *Rethinking Causality, Complexity and Evidence for the Unique Patient: A CauseHealth Resource for Health Professionals and the Clinical Encounter*, Dordrecht: Springer.

Anscombe, G. E. M. (1971) 'Causality and determination', in *Metaphysics and the Philosophy of Mind*, Oxford: Blackwell, 1981, pp. 133–47.

Beebee, H. (2000) 'The non-governing conception of laws of nature', *Philosophy and Phenomenological Research*, 61: 571–94.

Beebee, H. (2006) 'Does anything hold the universe together?', *Synthese*, 149: 509–33.

Bird, A., Ellis, B., and Sankey, H. (eds) (2012) *Properties, Powers and Structures: Issues in the Metaphysics of Realism*, London: Routledge.

Boobis, A., Budinsky, R., Collie, S., Crofton, K., Embry, M., Felter, S.,...Zaleski, R. (2011) 'Critical analysis of literature on low-dose synergy for use in screening chemical mixtures for risk assessment', *Critical Reviews in Toxicology*, 41: 369–83.

Callahan, M. A., and Sexton, K. (2007) 'If cumulative risk assessment is the answer what is the question?', *Environmental Health Perspectives*, 115: 799–806.

Cartwright, N. (1983) *How the Laws of Physics Lie*, Oxford: Clarendon Press.

Cartwright, N. (1994) *Nature's Capacities and Their Measurement*, Oxford: Oxford University Press.

Cartwright, N. (1999) *The Dappled World: A Study of the Boundaries of Science*, Cambridge: Cambridge University Press.

Cartwright, N. (2004) 'Causation: One word, many things', *Philosophy of Science*, 71: 805–20.

Cartwright, N. (2007) *Hunting Causes and Using Them: Approaches in Philosophy and Economics*, Cambridge: Cambridge University Press.

Cartwright, N. (2020) 'X—Why trust science? Reliability, particularity and the tangle of science', *Proceedings of the Aristotelian Society*, 120: 237–52.

Cellini, F., Chesson, A., Colquhoun, I., Constable, A., Davies, H. V., Engel, K. H., Gatehouse, A. M., Kärenlampi, S., Kok, E. J., Leguay, J. J., Lehesranta, S., Noteborn, H. P., Pedersen, J., and Smith, M. (2004) 'Unintended effects and their detection in genetically modified crops', *Food Chemistry and Toxicology*, 42: 1089–125.

Checker, M. (2007) '"But I know it's true": Environmental risk assessment, justice, and anthropology', *Human Organization*, 66: 112–24.

Cohrs, R. J., Martin, T., Ghahramani, P., Bidaut, L., Higgins, P. J., and Shahzad, A. (2015) 'Translational medicine definition by the European Society for Translational Medicine', *New Horizons in Translational Medicine*, 2: 86–8.

Collins, J. D., Hall, E. J., and Paul, L. A. (eds) (2004) *Causation and Counterfactuals*, Cambridge, MA: MIT Press.

Dally, A. (1998) 'Thalidomide: Was the tragedy preventable?', *The Lancet*, 351: 1197–9.

Dendena, B., and Corsi, S. (2015) 'The environmental and social impact assessment: A further step towards an integrated assessment process', *Journal of Cleaner Production*, 108: 965–77.

Dowe, P. (2000) *Physical Causation*, New York: Cambridge University Press.

Dumsday, T. (2019) *Dispositionalism and the Metaphysics of Science*, Cambridge: Cambridge University Press.

Dupré, J. (1981) 'Natural kinds and biological taxa', *Philosophical Review*, 90: 66–90.

Dupré, J. (1993) *The Disorder of Things: Metaphysical Foundations of the Disunity of Science*, Cambridge, MA: Harvard University Press.

Dupré, J. (2007) *The Constituents of Life*, The Spinoza Lectures, Assen: Van Gorcum.

Dupré, J., and Nicholson, D. J. (2018) 'A manifesto for a processual philosophy of biology', in D. J. Nicholson and J. Dupré (eds), *Everything Flows: Towards a Processual Philosophy of Biology*, Oxford: Oxford University Press, pp. 3–45.

Ellis, B. (2001) *Scientific Essentialism*, Cambridge: Cambridge University Press.

Festing, M. F. (2010) 'Inbred strains should replace outbred stocks in toxicology, safety testing, and drug development', *Toxicologic Pathology*, 38: 681–90.

Gillies, D. (2005) 'An action-related theory of causality', *British Journal for the Philosophy of Science*, 56: 823–42.

Gillies, D. (2018) *Causality, Probability, and Medicine*, London: Routledge.

Glennan, S. (1996) 'Mechanisms and the nature of causation', *Erkenntnis*, 44: 49–71.

Glennan, S. (2009) 'Mechanisms', in H. Beebee, C. Hitchcock, and P. Menzies (eds), *Oxford Handbook of Causation*, Oxford: Oxford University Press, pp. 315–25.

Greco, J., and Groff, R. (eds) (2013) *Powers and Capacities in Philosophy: The New Aristotelianism*, New York: Routledge.

Gurian-Sherman, D. (2009) 'Failure to yield: Evaluating the performance of genetically engineered crops', Union of Concerned Scientists, 14 April, https://www.ucsusa.org/resources/failure-yield-evaluating-performance-genetically-engineered-crops.

Guyatt, G., Cairns, J. A., Churchill, D., et al. (1992) 'Evidence-based medicine: A new approach to teaching the practice of medicine', *JAMA—Journal of the American Medical Association*, 268: 2420–5.

Hall, N. (2004) 'Two concepts of causation', in J. Collins, N. Hall, and L. Paul (eds), *Causation and Counterfactuals*, Cambridge, MA: MIT Press, pp. 225–76.

Harré, R., and Madden, E. H. (1975) *Causal Powers: A Theory of Natural Necessity*, Oxford: Blackwell.

Heil, J. (2004) 'Properties and powers', in D. Zimmerman (ed.), *Oxford Studies in Metaphysics*, vol. 1, New York: Oxford University Press, pp. 223–54.

Heil, J. (2017) 'Real modalities', in J. Jacobs (ed.), *Causal Powers*, Oxford: Oxford University Press, pp. 90–104.

Hicks, D. J. (2015) 'Epistemological depth in a GM crops controversy', *Studies in History and Philosophy of Biological and Biomedical Sciences*, 50: 1–12.

Hume, D. (1739) *A Treatise of Human Nature*, ed. L. A. Selby-Bigge, Oxford: Clarendon Press, 1888.

Hume, D. (1748) *An Enquiry Concerning Human Understanding*, ed. P. Millican, Oxford: Oxford University Press, 2007.

Illari, P. M. K. (2011) 'Mechanistic evidence: Disambiguating the Russo–Williamson thesis', *International Studies in the Philosophy of Science*, 25 (2): 139–57.

Ingthorsson, R. D. (2021) *A Powerful Particulars View of Causation*, London: Routledge.

Jacobs, J. (ed.) (2017) *Causal Powers*, Oxford: Oxford University Press.

Jakobsson, H. E., Rodríguez-Piñeiro, A. M., Schütte, A., Ermund, A., Boysen, P., Bemark, M., Sommer, F., Bäckhed, F., Hansson, G. C., and Johansson, M. E. (2015) 'The composition of the gut microbiota shapes the colon mucus barrier', *EMBO Reports*, 16: 164–77.

Kacew, S., Ruben, Z., and McConnell, R. F. (1995) 'Strain as a determinant factor in the differential responsiveness of rats to chemicals', *Toxicologic Pathology*, 23: 701–15.

Kerry, R., Eriksen, T. E., Lie, S. A. N., Mumford, S., and Anjum, R. L. (2012) 'Causation and evidence-based practice: An ontological review', *Journal of Evaluation in Clinical Practice: Special Issue on the Philosophy of Medicine and Health Care*, 18: 1006–12.

Kistler, M. (2006) *Causation and Laws of Nature*, New York: Routledge.

Kutach, D. (2007) 'The physical foundations of causation', in H. Price and R. Corry (eds), *Causation, Physics, and the Constitution of Reality: Russell's Republic Revisited*, Oxford: Oxford University Press, pp. 327–50.

Lewis, D. (1973) *Counterfactuals*, Oxford: Blackwell.

Lewis, D. (1973/1986) 'Causation', in *Philosophical Papers II*, Oxford: Oxford University Press 1986, pp. 159–213.

Mackie, J. L. (1980) *The Cement of the Universe: A Study of Causation*, New York: Oxford University Press.

Marmodoro, A. (ed.) (2010) *The Metaphysics of Powers: Their Grounding and Their Manifestations*, London: Routledge.

Marmodoro, A. (2016) 'Dispositional modality vis-à-vis conditional necessity', *Philosophical Investigations*, 39: 205–14.

Martin, C. B. (2008) *The Mind in Nature*, Oxford: Oxford University Press.

Mellor, D. H. (1971) *The Matter of Chance*, Cambridge: Cambridge University Press.

Mellor, D. H. (1995) *The Facts of Causation*, London: Routledge.

Molnar, G. (2003) *Powers: A Study in Metaphysics*, ed. S. Mumford, Oxford: Oxford University Press.

Moretto, A., Bachman, A., Boobis, A., Solomon, K. R., Pastoor, T. P., Wilks, M. F., and Embry, M. R. (2017) 'A framework for cumulative risk assessment in the 21st century', *Critical Reviews in Toxicology*, 47: 85–97.

Mumford, S. (1998) *Dispositions*, Oxford: Oxford University Press.

Mumford, S. (2004) *Laws in Nature*, Abingdon: Routledge.

Mumford, S. (2005) 'Laws and lawlessness', *Synthese*, 144: 397–413.

Mumford, S., and Anjum, R. L. (2011a) *Getting Causes from Powers*, Oxford: Oxford University Press.

Mumford, S., and Anjum, R. L. (2011b) 'Spoils to the vector: How to model causes if you are a realist about powers', *The Monist*, 94: 54–80.

Mumford, S., and Anjum, R. L. (2013) *Causation: A Very Short Introduction*, Oxford: Oxford University Press.

Nancy, A. and Eckardt, N. A. (2007) 'Systemic acquired acclimation to high light', *Plant Cell*, 19: 3838, doi: 10.1105/tpc.108.191210.

Norton, J. D. (2007) 'Causation as folk science', in H. Price and R. Corry (eds), *Causation, Physics, and the Constitution of Reality: Russell's Republic Revisited*, Oxford: Oxford University Press, pp. 11–44.

Peterson, C. H., Rice, S. D., Short, J. W., Esler, D., Bodkin, J. L., Ballachey, B. E., and Irons, D. B. (2003) 'Long-term ecosystem response to the Exxon Valdez oil spill', *Science*, 302: 2082–6.

Popper, K. (1990) *A World of Propensities*, Bristol: Thoemes.

Psillos, S. (2002) *Causation and Explanation*, Chesham: Acumen.

Richter, S. H., Garner, J. P., Zipser, B., Lewejohann, L., Sachser, N., Touma, C., Schindler, B., Chourbaji, S., Brandwein, C., Gass, P., and van Stipdonk, N. (2011) 'Effect of population heterogenization on the reproducibility of mouse behavior: A multi-laboratory study', *PLoS One*, 6: e16461.

Rocca, E., and Andersen, F. (2017) 'How biological background assumptions influence scientific risk evaluation of stacked genetically modified plants: An analysis of research hypotheses and argumentations', *Life Sciences, Society and Policy*, 13, online first: doi: 10.1186/s40504-017-0057-7.

Rocca, E. and Anjum, R. L. (2019) 'Why causal predictions fail: An example from oil contamination', *Ethics, Policy and Environment*, 22: 197–213.

Rocca, E. and Anjum, R. L. (2020) 'Causal evidence and dispositions in medicine and public health', *International Journal of Environmental Research and Public Health*, 17: 1813.

Rocca, E., Anjum, R. L., and Mumford, S. (2020) 'Causal insights from failure', in A. La Caze and B. Osimani (eds), *Uncertainty in Pharmacology: Epistemology, Methods and Decisions*, Cham: Springer, pp. 39–57.

Sackett, D. L. and Rosenberg, W. M. (1995) 'The need for evidence-based medicine', *Journal of the Royal Society of Medicine*, 88: 6204.

Sackett, D. L., Rosenberg, W. M., Gray, J. M., Haynes, R. B., and Richardson, W. S. (1996) 'Evidence based medicine: What it is and what it isn't', *British Medical Journal*, 312: 71.

Sawyer, S., and Loja, N. (2015) 'Crude contamination: Law, science, and indeterminacy in Ecuador and beyond', in H. Appel, A. Mason, and M. Watts (eds.), *Subterranean Estates: Life Worlds of Oil and Gas*, New York: Cornell University Press, pp. 126–46.

Sexton, K. (2015) 'Cumulative health risk assessment: Finding new ideas and escaping from the old ones, *Human and Ecological Risk Assessment: An International Journal*, 21 (4), 934–51, doi: 10.1080/10807039.2014.946346.

Seyhan, A. A. (2019) 'Lost in translation: The valley of death across preclinical and clinical divide—identification of problems and overcoming obstacles', *Translational Medicine Communications*, 4 (18), doi: 10.1186/s41231-019-0050-7.

Solomon, K. R., Wilks, M. F., Bachman, A., Boobis, A., Moretto, A., Pastoor, T. P.,... Embry, M. R. (2016) 'Problem formulation for risk assessment of combined exposures to chemicals and other stressors in humans', *Critical Reviews in Toxicology*, 46: 835–44.

Suppes, P. (1970) *A Probabilistic Theory of Causality*, Amsterdam: North-Holland.

Twerdok, L. (1999) 'Development of toxicity criteria for petroleum hydrocarbon fractions in the petroleum hydrocarbon criteria working group approach for risk-based management of total petroleum hydrocarbons in soils', *Drug and Chemical Toxicology*, 22: 275–91.

US EPA (2003) *Framework for Cumulative Risk Assessment.* PA/630/P-02/001A. Washington, DC: US EPA, Risk Assessment Forum, Office of Research and Development.

Vargesson, N. (2013) 'Thalidomide embryopathy: An enigmatic challenge', *ISRN Developmental Biology*, doi: 10.1155/2013/241016.

Vetter, B. (2015) *Potentiality: From Dispositions to Modality*, Oxford: Oxford University Press.

Wickson, F., and Wynne, B. (2012) 'Ethics of science for policy in the environmental governance of biotechnology', *Ethics, Policy and Environment*, 15: 321–40.

Williams, P. R. D., Dotson, G. S., and Maier, A. (2012) 'Cumulative risk assessment (CRA): Transforming the way we assess health risks', *Environmental Science and Technology*, 46: 10868–74.

Woodward, J. (2003) *Making Things Happen: A Theory of Causal Explanation*, Oxford: Oxford University Press.

6

Causal Efficacy

A Comparison of Rival Views

R. D. Ingthorsson

1. Introduction

The aim of this chapter is to critically discuss four rival views about how to best make sense of causation as a real mind-independent feature of the world that involves the production of changes through the exertion of influence of something on something else. Is causation a question of (i) transmission of conserved quantities (*transmission accounts*), (ii) activities of the parts of mechanisms (*mechanistic accounts*), (iii) mutual manifestation of powers (*powers-based accounts*), or (iv) reciprocal action between powerful particulars (*powerful particulars account*)? I argue in favour of (iv). Note that in this chapter I will use the term 'interaction' to denote any phenomenon where something exerts an influence on anything else, even though I will argue that the scientific understanding of interactions, the one I promote, is of *reciprocal action* between two entities, notably that whenever any entity *A* exerts any kind of influence on any other entity *B*, *B* will at the same time exert the same kind of influence on *A*, and to the same magnitude.

It bears to mention that transmission and mechanistic accounts are not always counted as causal realist accounts—that is, as treating causation as a mind-independently real phenomenon—but only as attempts to clarify causal reasoning. This is partly because some proponents explicitly take that stance, and partly because these accounts originate in the empiricist tradition of accounting for everything in terms of what can be observed and are therefore unreflectively grouped with neo-Humean accounts. Indeed, their proponents share an aversion to the postulation of anything popularly regarded as 'unobservable' such as powers. However, I believe the thinkers mentioned in this chapter do treat causation as a mind-independent and real phenomenon and therefore can count as causal realists.

I will first briefly sketch the core ideas of the different accounts before ventilating my concerns about each view, first in general terms and then with respect to the way they explain two kinds of causal phenomenon: (i) collision between billiard balls and (ii) how water dissolves salt. It will be argued that transmission, mechanistic, and powers-based accounts are, to varying degree, incompatible with

R. D. Ingthorsson, *Causal Efficacy: A Comparison of Rival Views* In: *Alternative Approaches to Causation: Beyond Difference-making and Mechanism.* Edited by: Yafeng Shan, Oxford University Press.
© Oxford University Press 2024. DOI: 10.1093/oso/9780192863485.003.0006

certain scientifically established facts about the behaviour of physical entities, in a way that the powerful particulars view is not.

2. Transmission Accounts

I will initially use Wesley Salmon's account of causation in terms of *causal processes, propagation,* and *production* (Salmon 1980, 1984) to illustrate the core idea of transmission accounts, since most contemporary accounts are influenced by it, such as Philip Kitcher (1989), Phil Dowe (1992), and Max Kistler (1998). Later, I will refer to Dowe (2009) for a more recent formal statement of the core tenets of transmission accounts.

It bears to mention that Salmon's account is sometimes described as 'neo-mechanistic' because it was originally a revival of the idea that explanation is about identifying the causal mechanisms that produce natural phenomena (Galavotti 2022). However, he ultimately moved towards a transmission account of causal mechanisms.

According to Salmon, the persistent entities that form the basis of most scientific ontologies—particles, molecules, cells, animals, planets, etc.—are to be understood as *causal processes*; they are causal because able to exert influence on each other, and they are processes because science reveals them to be continuously changing even though they may appear not to. Ultimately, everything above the level of elementary particles is either a continuously changing atom or made of atoms of that kind.

Salmon denies that talk of 'process' is a commitment to event ontology—that is, that entities are structured series of events. However, he doesn't offer an alternative analysis of process; he just points to examples of things that remain the same through continuous intrinsic change (1984: 139ff.). His rejection of event ontology is not a commitment to substance ontology either, but a rejection of *any* ontology that tries go beyond the observable. Of the transmission accounts that developed in Salmon's wake, some work with events as the relata of transmission (Kistler 1998) while others stick to Salmon-style causal processes (Dowe 2009).

Propagation refers to the way causal processes conserve their structure over time when not interacting with other processes and consequently conserve their ability to exert causal influence in any future interaction. *Production* refers to the way interactions between causal processes result in modifications in their structure and in their ability to exert causal influence.

Salmon resists commitment to causal powers for the same reason he resists commitment to event ontology; they are unobservable and therefore mysterious. Instead, he first uses the term 'marks' to refer to the observable properties of causal processes recognized by the sciences, and which play the same explanatory role that powers are meant to do (Salmon 1980). Later, due to the criticism from Kitcher (1989) and Dowe (1992), Salmon adopted the view that changes are produced in interacting causal processes because one transmits a conserved quantity to the other.

Salmon's view can be illustrated concretely by considering a batter hitting a baseball with a bat in such a way that it leaves the pitch and breaks a window in a neighbouring house. The bat and baseball are causal processes propagating the conserved quantity of momentum. On contact, a conserved quantity of momentum is transmitted from bat to ball. The ball then preserves its momentum as it travels across the pitch, propagating it from point A to point B. When the ball interacts with a window it is in virtue of the preserved momentum that it breaks the window.

Let us now look at a more formal statements of transmission accounts; one event-based and another based on causal processes. Kistler presents the event-based account in what he calls a 'reduction statement':

(T) Two events a and b are causally related in the sense that one is a cause of the other if and only if there exists a conserved quantity Q of which a particular amount P is transmitted between a and b.   (Kistler 1998: 1)

Unfortunately, the statement is not specific enough about the relevant sense of 'event' to distinguish causal and non-causal instances of transmission, and therefore doesn't distinguish between what Salmon calls propagation and production. If events are understood Kim-style—that is, as a particular a instantiating a property F at a time t—then conservation laws entail that any quantity present in any Kim-style event e_1 will be transmitted to the next event e_2 whether or not anything causal occurs between e_1 and e_2. For instance, a uniformly moving body will through any given temporal interval constitute a succession of distinct Kim-style events between which any conserved quantity possessed by any event e_i will be transmitted to a subsequent event e_j (see definition of 'world-line' below). To be fair, Kistler obviously assumes that everyone understands that (T) is only meant to apply to what happens in causal interactions between distinct entities. However, this needs to be spelled out explicitly to avoid objections by hair-splitting readers such as myself. More importantly, to make (T) more precise it seems necessary to introduce some term for the persistent particulars that are the constituents of events. There are other worries, but they apply just as much to the causal process version discussed below.

The causal process version specifies that while a causal process is continuously transmitting conserved quantities (really, *propagating* them), it is only when something is transmitted between distinct processes to *produce* change that we can talk about causal interactions. This position can be summed up in terms of three claims (for reference, see Dowe 2009: 219):

Causal process: an object that possesses a conserved quantity.

World-line: a spacetime trajectory of a causal process in which it propagates a non-zero amount of a conserved quantity between every spacetime point of that trajectory.

Causal interaction/production: an intersection of world lines A and B ($A \neq B$) that involves exchange of a conserved quantity producing a modification in A and B.

Dowe's causal process version of the transmission account resolves the worries I raised about the event-based version. However, there are other problems, most of which are also present in mechanistic and powers-based accounts. It is important for what comes later that, while an 'exchange' implies a certain kind of reciprocity between A and B—both change in proportional, albeit different ways—the reciprocity is always assumed to involve the loss of a quantity by A, which B gains; exchange goes one way from A to B. It is this unidirectionality of the exchange that is meant to constitute the direction of causation.

3. Mechanistic Accounts

The core idea of mechanistic accounts is that causation is the activities of compound parts of organized wholes that produce changes in either whole and/or parts. It is in fact a requirement that causes and effects *must* be connected by mechanisms (Glennan 2017: 145). There are some disagreements to be found between accounts, say, between the 'mechanism first' approach favoured by Stuart Glennan in earlier works (2009) and the 'activities first' approach advocated by Machamer (2004) and Bogen (2008). However, I don't think it is of any consequence for this chapter to confine the discussion to the most recent systematic presentation of the mechanistic philosophy found in Glennan's book *The New Mechanistic Philosophy* (2017). Indeed, the influences from Machamer and Bogen, as well as from William Bechtel and Adele Abrahamsen (2005), Carl Craver (2007), and Lindley Darden (2008), just to name a few, are clear to see. I will refer to this view as 'NMP'.

NMP differs from the empiricist and reductive accounts of causation it is sometimes associated with, in rejecting the idea that causes and effects can be understood as Kim-style events. NMP stresses the *processual* nature of causation as a natural consequence of taking causal production to consist in the *exertion* of influence between particulars, and that this exertion must be understood as an *activity* that cannot fit in an instant (Glennan 2017: 177). To be more precise, the causally relevant sense of 'event' is of particulars (plural) doing something *to each other* since a cause is never the activity of just one particular but an interaction between parts of a mechanism. This removes the need to further distinguish between causal and non-causal events.

In more detail, Glennan suggests we should understand causation in terms of *constitutive, precipitating,* and *chained* production. Constitutive production is what happens as a result of interactions between parts of a mechanism, and which can only be understood in terms of the activities of the whole mechanism. Glennan uses the example of searing a steak to illustrate. When a steak is put on a hot skillet, energy is transferred to the steak (which is precipitating production) until the temperature of the steak rises above a certain limit (+140°C). At that point the constituents of the steak—various sugars and amino acids—react with each other in what is known as the *Maillard reaction* to produce a variety of molecules responsible for a range of pleasant flavours. So, the activities of the parts of the steak produce in those very parts a change to turn raw meat to seared delicacy.

Precipitating production is 'the way in which one or more events produce another event...by creating start-up conditions for a different mechanism' (2017: 182). The skillet transferring heat to the steak is an example of an event that creates the start-up conditions for the Maillard reaction in the steak.

Finally, *chained production* is a connection between events separated by a chain of intermediary precipitating events which in turn involve chains of constitutive production. We might say the connection between someone buying a steak and eating it, boils down to a series of precipitating events comprising grabbing the steak at the store, putting it in the carrier bag, carrying it home, putting it onto the skillet, searing it, placing it on the plate, and finally eating it.

There are certain points I puzzle over, regarding the distinctions between constitutive and precipitating production and their relationship to each other. Chained production, on the other hand, is straightforwardly understood as a connection between two non-adjacent events consisting of a chain of precipitating production, but precipitating production is not obviously constituted by constitutive production. Constitutive and precipitating phenomena, we are told, mark the difference between 'mechanisms that produce phenomena (non-constitutive) and mechanisms that underlie phenomena (constitutive)' (Glennan 2017: 109). This could mean that the heat transmitted from skillet to steak produces the phenomenon of Maillard reaction, while the interactions between the constituents of the steak underlie the Maillard reaction. In that case, I would have guessed precipitating production would correspond to what Salmon calls 'production'— that is, an interaction between two distinct causal processes which produces a modification in them (skillet cools down, steak heats up)—while constitutive production would correspond to the interactions that constitute the changes internally to each process when two processes interact; something Salmon does not have a separate name for other than simply 'modification'. However, Glennan says that constitutive production corresponds to Salmon's production and offers such examples as hammering a nail and a horse pulling a cart (Glennan 2017: 184). To my mind, hammering a nail is in the same category of phenomena as skillet searing a steak. In both cases we arguably have two separate and well-defined wholes interacting with each other—hammer and nail+board (skillet and steak)—which produces a change in each other but neither whole appears to underlie the changes in the other, not in the same way the constituents of the steak underlie the Maillard reaction. Similarly, the hammer is not an underlying component of the phenomenon of the nail being driven into the board, but it is something that contributes to that change.

What I suspect we are seeing here is the difficulty of applying a theoretical model to a many-layered and thoroughly complex reality. There is a problem in demarcating clearly between mechanisms on roughly the same level, and the many layers of mechanism within any one of them. Is hammer and nail and board separate mechanisms or are they all united by the person wielding the hammer and steadying the nail and board? And underlying the amino acids and fats that underlie the Maillard reaction are the atoms of the amino acids and fats whose interactions also must be constitutive of the wholes of which they are parts.

NMP and transmission accounts are similar in that NMP admits (or can admit) that transmission of conserved quantities is one of the ways we can understand what goes on in causal interactions, as for instance when energy flows from skillet to steak. NMP could even admit that in the domain of particle interactions, transmission of conserved quantities may possibly be all we need to understand what is going on. However, NMP claims to offer a more general model that can be applied to other scientific disciplines that do not operate with conserved quantities. Another difference between transmission and NMP accounts is that the former treats a mechanism as a nexus between distinct entities, whereas the latter treats a mechanism as a complex system. I take these differences to be large enough to treat transmission and mechanistic accounts as distinct theories despite the similarities.

Finally, like transmission accounts, NMP tends to characterize the influence exerted between distinct objects in precipitating production as unidirectional (from skillet to steak). However, the assumption of unidirectionality is much less pronounced in NMP, especially when it comes to constitutive production where it is not at all clear that interactions are assumed to be unidirectional.

4. Powers-Based Accounts

Powers-based accounts (e.g. Martin 1997; Ellis 2001; Molnar 2003; Mumford and Anjum 2011; Heil 2012; Marmodoro 2017) are more heterogeneous than transmission and mechanistic accounts, mostly because of a disagreement about the nature of powers. *Dispositional essentialists* accept the categorical/dispositional distinction—a distinction firmly rooted in the empiricist tradition—and thus say there is a difference between the properties that determine the qualitative state of the object at any given time—that is, its *qualities*—and the properties that determine what the object is liable to do if certain conditions arise—that is, its *powers/dispositions* (e.g. see Ellis 2001). Proponents of the *identity theory* reject the distinction and insist that the very same properties that determine the object's qualitative state also determine its ability to affect and be affected; they are *powerful qualities* (e.g. see Ingthorsson 2013).

Dispositional essentialists (e.g. Ellis 2001; Molnar 2003) tend to think of causation as a matter of pure potentialities (or unmanifested powers) being triggered by stimuli to transition from potentiality to actuality, and thereby manifesting some qualitative state: the *manifestation*. Identity theorists (Martin 1997; Heil 2012; Ingthorsson 2013) tend to think of causation as a matter of two powerful qualities mutually modifying their bearers from one powerful qualitative state to another powerful qualitative state: something called a *mutual manifestation*. There is a greater heterogeneity within each view, but for the purposes of this discussion I see no reason to delve into that. The two conceptions presented here, of how something is manifested, are so close that dispositional essentialists have found it easy to identify their manifestations with mutual manifestations, under the assumption that their manifestations are also a joint product of power and stimuli.

Now, most powers-based accounts, regardless of the differences mentioned, accept something equivalent to the distinction between *active* and *passive* powers that we find in Aristotle, the Stoics, the Scholastics, and the natural philosophy of the early Enlightenment—that is, between the ability to exert an influence on other objects/powers (active) vs. the ability to change in response to an external influence (passive). Indeed, they will represent the direction of causation as the direction of the influence flowing from the object with the active power to the object with the passive power. A dispositional essentialist will say that a ball at rest can potentially move (it has the power to move while it is not actually moving), and that this potentiality will become an actuality when influenced by a ball in motion colliding with the ball at rest. The resulting motion will be a mutual manifestation of the active and passive potencies. Most identity theorist will say that what we perceive as a ball at rest is really a ball with an actual and determinate momentum, p, which is understood as a state of motion. If such a ball, a, with momentum p_a, collides with another ball, b, with momentum p_b, the two balls will mutually modify each other in accordance with the laws of motion, resulting in a transition of a from p_a to p_{a^*}, and of b from p_b to p_{b^*}. However, despite differences in the manner in which dispositional essentialists and identity theorists characterize the collision between billiard balls, they will agree that one of the balls exerts an influence on the other, while the latter receives the influence. Like transmission accounts, most powers-based accounts characterize causal influence as unidirectional.

It is now time to introduce the powerful particulars view, but it is best to first point out what I think is problematic in the views already presented, because the powerful particulars view is best understood as an attempt to overcome those problems.

5. Influence: Unidirectional or Reciprocal?

Transmission, mechanistic, and powers-based accounts agree that causation is the exertion of influence of something on something else but disagree on the nature of the 'something' that exerts influence and about the nature of the influence being exerted. Most importantly, they all tend to treat influence as *unidirectional*. This is an overgeneralization, as already hinted at when presenting NMP. Indeed, it will emerge that I am an identity theorist rejecting the unidirectional influence. But it is true enough for purposes of presentation. I'll qualify it later with respect to variations within each family of views.

In taking influence to be unidirectional, the accounts under scrutiny are implicitly endorsing an understanding of influence that has been a part of the causal realist tradition since Aristotle wrote that 'whenever the potential active and the potentially affected items are associated in conditions propitious to the potentiality, the former must of necessity act and the latter must of necessity be affected' (1998: 9, 5; 264). Indeed, we see a similar agreement about the unidirectionality of influence in the Aristotelian, atomistic, Stoic, Scholastic, and natural philosophy of the early modern period as the one we see today. They all share certain core notions, that together make up what I have elsewhere called *the standard view* (for further details, see Ingthorsson 2021: 45ff.). The idea is that a new state is produced when an already existing entity, or complex of entities, changes due to an influence external to that entity, one without which the change would not have occurred, and the new state never exist. The external influence comes from an entity possessing powers that allow it to influence other entities (active), and the entity upon which it acts possesses the power to receive the influence and to change in some particular way (passive).

It is true that if we limit the discussion to the kind of examples that philosophers typically consider, the unidirectionality of influence and the distinction between active and passive entities appear to make good sense. Philosophers talk about billiard balls in motion acting on a ball at rest (Hume 1748: §36); a lead ball dropped onto a pillow produces a hollow (Kant 1787: A203); a locomotive pulls a truck (Taylor 1973: 35); a baseball is hit by a bat to fly across the pitch (Salmon 1980: 50). However, as Mario Bunge first pointed out (1959: ch. 6), it should be recognized as a serious problem that modern science categorically rejects the reality of unidirectional action even in the apparently asymmetric cases that philosophers typically consider. It is instead insisted that all influence comes in the form of *reciprocal action*, or 'interaction' as the term is defined in classical physics.

To be sure, classical physics is in many ways an outmoded framework, but we need to be careful in dismissing every component. While the first and second laws of motion are known to fail in extreme situations, the core idea expressed in the third law is still believed to hold good even in quantum and relativistic physics (albeit adapted to fit more generally to conservation laws), and it is the third law that is the basis for the rejection of unidirectional action. The core idea of the third law can be generalized in the following way:

Whenever any object whatsoever exerts any influence whatsoever on any other object whatsoever, the latter exerts at the same time an influence of the same magnitude on the former, but in the opposite direction.

While classical mechanics offer a fairly accessible treatment of the notion of reciprocal action, a full understanding of the concept still requires something of a shift of perspective. The notion is most clearly expressed in the third law of motion which says that the force by which object 1 acts on object 2 is equal to the oppositely directed force by which object 2 acts on object 1 ($F_{1\text{on}2} = -F_{2\text{on}1}$). However, because the prevailing understanding that influence is unidirectional has been such an integral part of the popular and philosophical understanding of causation for so long, the third law has always been widely misunderstood. Indeed, Newton's unfortunate decision to use the words *action* and *reaction* when explaining the law in plain English contributed to that misunderstanding. Let me explain.

The correct understanding of reciprocal action is not that of a phenomenon *composed* of two different kinds of action, of which one (the action) gives rise to or provokes the other (the reaction)—as in two tennis-players returning each other's strokes—but of a *single* phenomenon of mutual influence occurring simultaneously between two objects, such as when bat collides with baseball. In Bunge's words, 'physical action and reaction are, then, two aspects of a single phenomenon of reciprocal action' (1959: 153). Or, as physicists Resnick, Halliday, and Krane put it: 'Any single force is only one aspect of a mutual interaction between two bodies' (2002: 83). In reciprocal actions, neither side has priority, and the terms 'action' and 'reaction' can be arbitrarily assigned to either.

However, researchers are not always interested in both sides of an interaction equally, and therefore often focus on the effect that matters to them, neglecting other outcomes. The point is stated beautifully by James Clerk Maxwell:

The mutual action between two portions of matter receives different names according to the aspect under which it is studied, and this aspect depends on the extent of the material system which forms the subject of our attention. If we take into account the whole phenomenon of the action between the two portions of matter, we call it Stress...But if...we confine our attention to one of the portions of matter, we see, as it were, only one side of the transaction—namely, that which affects the portion of matter under our consideration—and we call this aspect of the phenomenon, with reference to its effect, an External Force acting on that portion of matter. The other aspect of the stress is called the Reaction on the other portion of matter.   (Maxwell 1877: 26–7)

The lesson to be learnt is that even if it is recognized that interactions are perfectly reciprocal, they are often treated, for the sake of convenience, as if they were instances of unidirectional flow of influence from one portion of matter to the other.

However, even if it is a part of the amassed scientific knowledge that interactions are reciprocal, misunderstandings are pretty common even among professional physicists—at least if we are to believe physics educators Steinberg, Brown, and Clement (1990) as well as Hellingman (1992). The main misconception they identify is the following:

> The 'inappropriate conception' (as Steinberg et al. call it) here at stake is the same as in Newton's time: action and reaction are conceived as separate agents instead of as two sides of one interaction. Since they are conceived this way and since there are always two interacting bodies in producing a force, the suggestion of an action belonging to one body and a reaction belonging to another is virtually inescapable.   (Hellingman 1992: 112)

Anyone teaching classical mechanics faces the challenge of how to prevent students from falling under the spell of the 'virtually inescapable suggestion'.

But what are the philosophically interesting consequences that follow from the established fact that interactions really are reciprocal and not unidirectional? Like Bunge, I think we must accept that the polarization of interacting entities into an Agent that exerts an influence and a Patient that suffers a change upon receiving an influence 'is ontologically inadequate' (Bunge 1959: 170–1). The reciprocity of interactions shows that there are no strictly passive entities—those who only receive influence but do not themselves influence other things—nor are there entities who influence other things without being themselves equally affected. Furthermore, since the mutual exertion of influence occurs simultaneously in equal magnitude in opposite directions between two objects, there is no way to give priority to one or the other as chiefly responsible for the outcome or even for initiating the interaction. It is because of this that the terms 'action' and 'reaction' are considered arbitrary and why 'we are free to consider either of them as the force or the counterforce' (Hertz 1956: 185). If the cause really comes first, we would not be free to do this. Ultimately, the conclusion is that in so far as any causal account assumes unidirectionality of influence, it is based on an empirical falsity in light of the current understanding of physical interactions.

There are two worries I expect many readers will find it difficult to let go of. One arises from the fact that physical interactions are indeed so often described and understood in terms of transmission of quantities, even in physics. Is physics both right and wrong, but in different ways, about how we should understand interactions? The other arises from the widespread use of the terms 'mutual' and 'reciprocal' in the literature about powers-based accounts. How exactly is reciprocal action as defined by physics different from mutual manifestations of disposition-partners as defined by powers-based accounts?

To answer the first question, then yes, physicists often apply the strategy Maxwell describes, notably to treat interactions as if only one side of the transaction mattered, such as when describing interactions in terms of the transmission of conserved quantities. The inadequacy of this kind of characterization comes out most clearly in the fact that while they often function fine for the intents and purposes of a given occasion, they fail to generalize: they don't work for symmetrical interactions. When we turn our attention to the whole phenomenon, the transmission account only half-explains. As Maxwell points out, this is a choice of convenience we can safely make when the other side of the 'transaction' doesn't matter to us.

Consider a cue ball striking the black eight at a slight angle, pushing the black eight into a corner pocket while the cue ball continues in a new direction to end in position for the next shot. It seems easy to explain this by the transition of a quantity from the cue ball to the other, but then we are neglecting the details about what makes the cue ball change its trajectory in the way it does. Can we explain its change of trajectory merely by its loss of a quantity without bringing in any notion of an influence being exerted by the black eight on the cue ball? I at least find it difficult to entirely replace talk of forces with talk of transmitted quantities. You would have to make it so that the loss of the momentum p by the cue ball not only takes the role of the force usually meant to act on the black eight but also of the force presumed to act on the cue ball to change its trajectory. This is difficult to do even in asymmetrical interactions, but even harder in symmetrical interactions, say, when two cue balls move in opposite directions with the same velocity and collide head on. The outcome cannot be explained only in terms of something being lost by A which is gained by B. At least it must involve both A and B losing and gaining something.

An interesting example to consider is the common practice of explaining particle interactions by the transference of a virtual particle from one to the other, represented by vertices in Feynman diagrams. This is a way of conceptualizing what happens as if something carries a quantity from A to B. But when we consider that the term 'virtual particle' is not generally believed to denote actual particles in motion (Jaeger 2021), the imagery of a transmission of something from one thing to the other appears to be misleading. Very simplified, a virtual particle is a technical notion used to signify the presence of certain theoretically calculated quantities considered to mediate interactions between real particles, and which obtain within time limits that are too narrow for anything to be observed. The idea is that the calculated quantities are of the kind that particles may have but it just isn't clear there are any, wherefore there is talk of a 'virtual' particle. The quantities kind of add up to something that might belong to a particle but may better be said to belong to a quantum field. However, if we were to assume there is a particle that actually moves between the two colliding particles, then in light of the reciprocity of interactions stipulated by the third law, we would have to ask in what direction the particle is moving—from particle A to B, or vice versa—and whether the assumption of it moving one way, or the other, would really explain the whole interaction or only one side of the 'transaction'. Again, it is easier to think of asymmetrical interactions in terms of transmission of a virtual particle from A to B than it is to think of symmetrical interactions in that way.

Turning now to the question of whether powers-based accounts already operate with an understanding of reciprocity when using terms like 'mutual manifestation of reciprocal disposition-partners', I think clearly not. According to physics, interactions are reciprocal in the sense that any two interacting things simultaneously influence each other and to the same magnitude. Powers-based accounts describe powers as reciprocal even when it is assumed that one thing influences and the other merely receives the influence. For them, to say that powers are 'reciprocal' is only an admission of the fact that two powers are always involved, and that both are considered equally important for the production of the outcome, but the outcome is typically only the change produced in the passive recipient. An example would be that the power of fire to heat and the ability of a hand to be heated both contribute to the end result of a hot hand. This is the traditional way of distinguishing between active and passive powers as well as between Agents and Patients. The point is that if we accept what physics takes to be an established fact—that all interacting objects influence each other—it is no longer possible to argue that the distinctions between active vs. passive powers and Agent vs. Patient are based on the unidirectionality of causal influence. This realization is the starting point of the powerful particulars view.

6. Powerful Particulars Account

I have now presented the core ideas of what must surely count as the main causal realist accounts in recent times, the most serious objection to them all, and some worries specific to each view. It is time to add my preferred account of causation in terms of reciprocal action between powerful particulars into the mix. It first appeared in the chapter 'Causal Production as Interaction' (Ingthorsson 2002) but again in more developed form in a monograph (Ingthorsson 2021). The basic idea is that a single conception of causal interactions can account—in one and the same way—for (a) *changes* produced by interactions between previously unconnected entities, whether simple or compound, (b) *composition* of compound unities, (c) *persistence* of such wholes over time, and (d) that changes *internal* to a whole are just as causal as the changes resulting from interactions between wholes. It does so in a manner that easily relates to the theories and findings of the empirical sciences, in two ways. First, it is transparent how the philosophical account and the theories and data of the sciences offer complementary accounts on different levels; philosophy providing a general model applicable to all the specific phenomena, different scientific disciplines providing the details about each particular phenomenon. Second, it is clear how future developments in the sciences could falsify the general philosophical account. For instance, if physics ever finds truly asymmetrical interactions that violate the third law, my view is falsified.

Accounts of causation in terms of interactions between powerful particulars is no novelty. They are found in Aristotle, the Stoics, the atomists, the Scholastics, and the natural philosophers of the early Enlightenment (for further discussion, see Ingthorsson 2021: ch. 3.7). In the twentieth century we find Dorothy Emmet (1985) and Ingvar Johansson (1989) defending this view. However, with the exception of Johansson's 'action by mixture' view, these earlier versions either assumed the unidirectionality of influence or construed the reciprocity of inter-actions more in the form of mutual manifestations and therefore are subject to the same kind of criticism levelled at powers-based accounts. More recently—indeed, while making final revisions of this chapter—Andrew Newman has independently offered an account of the constitution of objects in terms of interactions between their constituent parts, one that appeals to our scientific understanding of how particles form atoms (Newman 2022). Newman doesn't address water dissolving salt, but his understanding of the collision of billiard balls (2022: 307–8) agrees exactly with the analysis I offer below, as well as with my earlier analysis of a brick hitting a window (Ingthorsson 2002: §3).

My powerful particulars view aims to offer a unified explanation of what Glennan calls precipitative and constitutive production (2017: ch. 7), what Craver and Darden think is needed to 'maintain' certain structures (2013: 65–6), what explains the production and destruction of the kind of entities that Salmon calls 'causal processes' as well as the unified wholes that NMP would denote as 'mechanisms'. Bear in mind that I only describe my view in enough detail to make comparisons between views meaningful (for full details, see Ingthorsson 2021).

The powerful particulars view bears very little resemblance to transmission accounts, although it is meant to fit well to the scientific conception of the world as ultimately constituted by particles carrying properties of the kind described as conserved quantities. It resembles powers-based accounts, some more than others, to the degree that I think the best way to explain the behaviour of the entities that are parts of organized wholes (both why they constitute wholes and how they and the wholes behave in interactions) is to attribute powers to them. However, I prefer the conception of powerful qualities and I like powerful qualities to correspond to the natural properties that the sciences think we have good reason to think are real; I don't think powers come in addition to the natural properties, as something an object has 'in virtue of' its other qualities (Ingthorsson 2013).

This still leaves place for plenty of emergent properties—that is, properties of wholes that cannot be reduced to the sum of the properties of the parts.

I also prefer to think that it is the particulars that bear the properties that exert influence on each other, and which change, because I cannot make much sense of the idea that powers act on other powers to change the powers. Do we say a force makes a quality of velocity speed up from 10 to 16 km/hr? No, we say a force exerted between bodies makes a body accelerate from 10 to 16 km/hr. My reasons for preferring a 'particulars first' ontology are closely connected to my views on time, change, and identity, but this is not the place to elaborate (for details, see Ingthorsson 2002, 2016).

I am not convinced about the legitimacy of the distinction between active and passive powers or between agents and patients, at least not as they are drawn today: that is, as power to affect (active) vs. power to be affected (passive), whose possession determines whether an object is active (an agent) or passive (a patient). I might be persuaded to think there is a real distinction between active and passive powers of a single object, but not that they determine that some objects are active while others are passive. The only distinction between agents and patients that I am prepared to admit at present is the one between intentional agents and inanimate objects. If there is anything that can truly initiate an action, it would be an intentional agent.

I think my view resembles NMP the most, and if its proponents could be persuaded to think that the concept of powers need not be of unobservable mysterious potencies but simply of powerful qualities such as spin, charge, momentum, valency, etc.—the kind of properties that are ubiquitous in the ontology of all the natural sciences—then I think they could warm up to a powerful particulars view. In the end, the resemblance is mostly that we seem to identify the same kind of phenomena as belonging to the class of relevant causal phenomena of which we need to develop a unified view, and this I believe is a consequence of the shared idea that the theories and findings of the empirical sciences are an important input, and that an agreement between the philosophical and scientific images of the world is desirable.

It is relevant to note that the powerful particulars account arose from the realization that for more than 333 years the philosophical implications of one of the most significant results of classical physics has either not registered as relevant in the minds of philosophers of causation or been misunderstood and dismissed. Mario Bunge (1959) is a notable exception. I am talking about the result that there are no unidirectional actions, only reciprocal interactions. To me it appeared foolish to challenge the validity of this result, because it would require me to show that our current physics is fundamentally wrong. It seemed more fruitful to explore the consequences of accepting the result as true (as already noted, I treat it as a provisional and falsifiable truth). Bunge had already argued convincingly that the reciprocity of interactions demonstrates that the prevailing idea must be wrong that causes are active objects or events that exert influence on some passive recipient, and that effects are the changes suffered by the recipient. This idea can only be considered an approximation (Bunge (1959: 151ff.) calls it the 'causal approximation') which in many or most cases might be good enough given our particular explanatory interests, pretty much in the way Maxwell describes. However, Bunge comes to the conclusion that an account of causation based on reciprocal action is equal to a reduction of the asymmetric and productive relation of causation, to a non-productive and symmetric functional relationship of the kind Russell (1912) favours. I disagree with this conclusion. Yes, interactions are symmetrical in the sense that two interacting objects influence each other simultaneously and to the same magnitude, but that influence produces a succession of states between which there is an asymmetric relation of one-sided existential dependence: one of producer to product.

Let me repeat that the kind of reciprocity we can see traces of in NMP and powers-based accounts is not drawn from the realization that all interactions are reciprocal. It rather boils down to a realization that even the scientific explanations that only pay attention to one side of an interaction—in a so-called 'causal approximation'—cannot entirely ignore the role of the perceived 'passive' recipient. To fully embrace the reciprocity of interactions as expressed by the third law, one need also embrace the idea that interactions are not composed of two actions belonging to two entities but are a single phenomenon of reciprocal action.

In the end, despite undoubtedly being a conceptual revolutionary, I think Bunge was not fully able to embrace the conclusion of his own argument (for details, see Ingthorsson 2021: ch. 4.5). He insists that interactions are a single phenomenon of mutual action, but then rejects interaction as a basic principle for causation because it would involve the error of singling out the 'action' as cause and the 'reaction' as effect and therefore characterize the relationship between cause and effect as symmetric, functional, and non-productive. I am still puzzled why he did not realize that his own argument suggested instead that we must accept that neither action nor reaction can count as cause, but that the interaction as a whole—being a single, unified phenomenon—must count as the cause of whatever changes it brings about. On this view, it is not the baseball hitting a window that causes a breakage; that is only half the story. The interaction causes not just a breakage but a change in the state of the ball such that we end up with a ball at rest in a pile of broken glass. Similarly, the mutual action between billiard balls, regardless of their initial state of motion, causes quantitatively proportional changes in the state of both balls. It takes an equal amount of work to stop a ball as to move a ball.

It is often objected that the assumed reciprocity of interactions fails to account for the asymmetry exhibited by many interactions. The window breaks, but the ball does not. The standard explanation to asymmetries of this kind is simply that since the initial states of the two things are different, then one and the same influence exerted on them will lead to different changes. If you hit a window with a hammer, it will break. If you hit a rubber ball in exactly the same way, it will not break. Instead, the hammer may bounce back to hit you in the face. The observed asymmetry is perfectly compatible with the reciprocity of interactions.

I will admit though that *if* a case is to be made for a distinction between active and passive powers and agents and patients—one that still respects the reciprocity of interactions—it is to argue that in many cases the consequences on either side of the interaction must be considered in some sense graver for one than the other, even though they are otherwise proportional. Davis Kuykendall (forthcoming) argues just that, developing an earlier suggestion from Anna Marmodoro (2017). For instance, Kuykendall (forthcoming: §4.1) suggests that enzymes, as biological catalysts, speed up chemical reactions without themselves being destroyed by the reaction. He argues in a similar vein, in §4.2 of the paper, that the interaction between H_2O and NaCl results in the destruction of NaCl but not H_2O. My answer is, first, that asymmetries of that kind do not violate the reciprocity of interactions; they can be explained by appeal to the differences in the initial states of the two interacting things. Second, to find a handful of asymmetrical examples does not justify the conclusion that causal interactions are generally asymmetrical. It can at best show that sometimes interactions bear signs of some kind of asymmetry, but not of the kind that contradicts the third law of motion.

According to the powerful particulars view, causation is best described in terms of reciprocal action, and the particular ways that entities can affect each other is determined by their intrinsic and powerful qualities; qualities whose determinate nature is best discovered by science. Indeed, science has found out that there are four different kinds of fundamental interaction, some of which are attractive while others are repulsive, some stronger than others, and they operate at different distances. We are talking about the four fundamental forces of nature. However, for all of them it is true—despite differences—that they exert their force reciprocally between interacting entities.

If causation is reciprocal action which sometimes comes in the form of attraction and sometimes as repulsion, the lesson I draw is that philosophy has only been looking at a subclass of causal phenomena, notably where previously unconnected entities suddenly interact, often with very disruptive consequences; bricks breaking windows, balls in motion disturbing balls at rest, etc. Philosophy has focused on this subclass because of the assumption that causation involves a unidirectional influence of something on something else and therefore essentially involves an *external* compulsion; and also, of course, because the interactions taking place between the component parts of objects could not be observed until quite recently. From the assumption that causation involves an external compulsion, it follows that anything happening inside an object, especially if it does not produce visible changes on the surface, is judged to be non-causal. Consider that the definition of spontaneous change is of a change that happens in the absence of an external compulsion. Even with the knowledge that only compound entities decay and knowing that compounds are held together by interactions between the component parts (which, on the hypothesis being considered, are causal), the idea that causation involves external compulsion has made it difficult to think of decay as a causal process. Hence it is called spontaneous decay. However, if causation is the phenomenon of reciprocal action, and such interactions occur within compound entities, then interactions between the parts of compound objects are also causal. Even if such interactions do not always produce visible disruption and change but instead appear only to maintain the inner structure of the entity, this is no reason to conclude that constitution and persistence are non-causal. Indeed, considering our current knowledge of the physical constitution of compounds, it is misleading to say that they *maintain* an inner structure, at least if you interpret 'maintain' as simply remaining the same. As far as I know, there are no interactions that do not involve some form of continuous change, even though many of them also preserve certain structures. All known compounds, beginning with protons and neutrons as compounds of quarks, are thoroughly dynamic entities in the sense that while they may stay the same structurally, they are still continuously changing on a more fine-grained level. Indeed, that change is what gives them their stability, which is why it makes sense to say that continuous reciprocal actions between the parts of a whole can *produce* stability, since that stability is dynamic. Just consider any atom of your choice. They are composed of a nucleus of protons and neutrons 'encircled' by a cloud of electrons. Every component part of that atom, as well as the atom as a whole, is continually changing. It follows that everything made up of such atoms is continually changing too. Consequently, such wholes are processes, if by 'process' we mean any entity for which change is essential.

I hope I have now said enough to make it possible to compare the four different accounts with respect to how well they handle two different concrete cases: (i) collisions between billiard balls and (ii) water dissolving salt.

7. Case Study I: Collisions between Billiard Balls

7.1 Transmission Accounts

The inadequacies of the transmission account when it comes to explaining collisions between billiard balls has already been drafted in some detail above, so I'll be brief but still add a couple of details. I said that transmission accounts appear to make intuitive sense in cases like 'ball in motion acts on ball at rest', but not when two identical billiard balls moving at the same speed in opposite directions collide head on; the account doesn't generalize to fit all the cases.

Furthermore, it is relevant to point out that it is only under the assumption that the balls are perfectly rigid, and when friction against the table is ignored, that transmission takes place only between the balls. In the real world of billiards—considering only symmetrical collisions—the balls take away from the interaction an equal share of conserved quantities, but some energy is dissipated as heat and as sound. We now have at least three directions in which conserved quantities flow. Which of these directions represents the direction of causation? To say it is the direction that matters most to the player is to decide to only look at one side of the story for anthropocentric reasons. My suggestion is that a deeper account of how interacting particulars exert an influence on each other is needed, one which explains why the conserved quantities are distributed/changed the way they are when particulars interact. A mere description of the actual exchange of quantities doesn't answer that question. If you look at the full range of cases, there is no single direction in which quantities flow between entities.

Traditionally, the concept of 'force' has served as the explanation of how interacting particulars influence each other. However, in the friction between reductionistic empiricism and anti-reductionist rationalism, the concept of force has been just as controversial as the notion of power. Empiricists like Hertz (1956) and Mach (1919) wanted to get rid of the concept. To them the notion of force was an unnecessary postulate based on a redundant inference from observed changes to some imagined invisible cause to those changes. They instead wanted to describe interactions between material entities merely in terms of the changes in the state of motion that they can be observed to suffer. Accordingly, the second law of motion should not really be understood as saying that in any interaction there is this special thing, a force, that is *proportional in magnitude* to the object's mass times the acceleration it suffers, but as stating an *identity* of force and change in state of motion. Really, the third law of motion ($F_{1on2} = -F_{2on1}$) can then just as well be expressed by a reconstructed two-way second law: $m_1 \times a_1 = m_2 \times a_2$. In plain English, when two material systems interact, the observed change in the state of motion of the first ($m_1 \times a_1$) is always equal in magnitude to the observed change in the state of motion of the second ($m_2 \times a_2$). But, having effectively removed the notion of force from the equation, we no longer have an explanation as to what it is that causes the changes in the two material systems; we just have a description. Why should we accept a reduction of this kind? The empiricist answer is that we otherwise must appeal to the mysterious notion of force.

Contemporary proponents of transmission should not have any qualms about the concept of force, in so far as they seek to ground their view in the notions already in use in the natural sciences. Appeals to force are ubiquitous in physics. However, to accept it, transmission theorists have to accept that there is a more fundamental feature to physical interactions than the transmission of conserved quantities, notably something (force/influence) that makes the quantities be transmitted in a particular way. Transmission accounts avoid commitment to the reality of forces and therefore end up being merely descriptive and in fact empirically inadequate when we consider the full range of interactions.

7.2 Mechanistic Accounts

The worries I have about mechanical explanations of billiard balls colliding are, first, that it isn't clear to me whether they would be treated as reciprocal (and therefore more like constitutive production) or unidirectional (and therefore more like precipitative production). The difficulty is partly to decide whether the change suffered by a ball in motion would be considered of 'lesser' importance than the change suffered by the ball at rest, say, because it matters more to the player that a ball at rest goes into a pocket. However, for an experienced player it is equally important to down a ball as it is to place the cue ball in position for the next shot. They must consider the effects on both balls equally. Is it the novice or experienced player who is best equipped to decide which side of the transaction matters? To my mind, the criteria for judging the direction of interactions in terms of 'importance' seem clearly anthropocentric and will be a case of deciding on the basis of the interest of individual players which side of the interaction they favour. At the very least, assuming the third law is valid, whether any change is 'lesser' in importance must be wholly unrelated to the purely physical magnitude of any change. We must then still accept the reciprocity of interactions but, like Kuykendall (forthcoming) and Marmodoro (2017), may perhaps look elsewhere for objective criteria for treating interactions as asymmetric. Kuykendall and Marmodoro's suggestion is that if A interacts with B with the result that B breaks but not A, then we have a kind of 'directedness'—perhaps not one that could be accounted for in terms of quantities, but still objectively real. However, the handful of examples that might plausibly be considered asymmetric in this sense would only show that some interactions are directed; we are short of a generalizable account of causation. I will have more to say about that when discussing water dissolving salt.

Second, I am uncertain of how to apply the notion of mechanism to collisions between billiard balls, especially if we are to follow the idea that every interaction is an activity between parts of a mechanism. How exactly are the billiard balls part of a mechanism? I am worried that the sense in which they are part of a mechanism is either anthropocentric or turns out to include the entire universe. To be sure, the balls and table are designed to allow the balls to roll in certain restricted ways, for the purposes of the game. We also have the cue and the players, but where do we draw the lines for this particular mechanism? Is it only the table, cue, balls, and players, or is it also the room, the building, the planet, and solar system? There doesn't seem to be any particular physical bond between balls, table, cue, and players that organizes them in any particular way in which they are not equally connected to the floor on which the table is standing, or the planet on which the building stands. The molecules of each individual ball are, however, clearly connected to each other in a way that the ball is not connected to the table. But such bonds do not obtain between the balls, or between them and the table. To be sure, the balls, table, cue, and players form a unity in our minds, but that would again introduce an unwanted anthropocentric feature. If proponents of NMP want the theory to be a contender among mind-independent theories of product-ive causation, they must provide an account of how billiard balls are parts of a mechanism that doesn't rely on human cognition.

The question perhaps ultimately is whether it is unnecessarily confining to require that every causal interaction be between parts of an already existing organized whole. Can we not talk about interactions between previously uncon-nected entities? If the case of billiard balls colliding is not persuasive enough, consider the random collision between two pieces of rock drifting aimlessly in space. How can we understand their interaction as an activity of the parts of an organized whole? And if we can't, is their collision not causal?

An alternative is to ask whether previously unconnected entities may achieve some kind of unity when the interaction starts, such that they become parts of an organized whole during the interaction even if they were not so connected before. I have suggested that this is in fact what happens in interactions (Ingthorsson 2021: ch. 4). An interaction, on the realist stance taken here, is a substantial connection between two or more entities. This is clearest in cases when two entities attract each other to form a compound, but the same is the case in repulsion. Indeed, once interactions are accepted as the basic mechanism of causation, we have already accepted that interacting entities are parts of an organized whole. In any interaction—even between previously unrelated entities—the entities achieve a substantial connection in virtue of the forces they exert on each other (attractive or repulsive) and then form a unity of parts acting on each other, albeit the unity sometimes is very short lived. Indeed, the general understanding of interactions as a unified phenomenon rather than a composite of two separate actions implicitly supports this understanding. Accordingly, two colliding billiard balls become a mechanism on contact and continue to be one as long as they are in contact. Indeed, on this view there really is no difference between precipitating and constitutive production, other than this: precipitating production is an interaction between previously unconnected entities, while constitutive production is an interaction between already con-nected parts of a whole. Indeed, I have suggested that reciprocal action allows us to understand not just the changes that happen inside a steak when it is being seared as being causal, but also the material constitution of the steak during periods when the interactions between its constituent parts only serve to main-tain the steak—for instance, while it is being carried home from the butchers (Ingthorsson 2021: ch. 6).

In sum, my worries about NMP revolve around an uncertainty in the application of the model to particular cases, an uncertainty that arises partly because it hasn't been designed to take an explicit stance on the problems I raise here and elsewhere concerning the age-old view that exertion of influence is unidirectional (Ingthorsson 2002, 2021), and partly because NMP does not have an account of what it is for entities to constitute a mechanism that seems to cover all putative cases of causal interactions. As far as I can see, if NMP were to embrace my account of causation as reciprocal interaction as the fundamental feature of causation, both these problems would be solved.

7.3 Powers-Based Accounts

Powers-based accounts are more heterogeneous than transmission and mechanistic accounts and therefore more difficult to present and criticize all in one go. One general worry already mentioned is that they tend to explain collisions, like transmission accounts, in terms of an exertion of influence of one ball on another which receives the influence, which is in conflict with the result that there are no unidirectional actions. Then there are worries connected to the specific conceptions of powers.

Dispositional essentialists characterize the powers of the balls in terms of potencies—that is, properties that are not instantiated by the object until the change is being manifested in a collision. Accordingly, the ball at rest can potentially move, and the moving ball can potentially make another ball move. My worry is how to reconcile this with the scientific picture. According to science, the billiard balls are solid and have a shape that together allows them to roll. More importantly, they have at any given time a quantifiable and directed momentum, which is represented as a property that the balls have before the collision, and which is what supposedly makes the balls able to influence each other. None of these properties that figure in the scientific explanation are pure potencies, but rather are actual and occurrent properties of the balls. If there were any pure potencies in that picture, they would have to be something in addition to these determinate and fully realized properties. The dispositional essentialist must postulate that in addition to, and independent from, a given ball's actual momentum, it has a potency to change its momentum p_1 to another momentum p_2 by going continuously through the intermediaries. But on this view, momentum is itself an inert property, but perhaps one from which emerges a power to change momentum.

My objection to dispositional essentialism is not that it is incoherent but that it is needlessly complicated and does not easily link up with the scientific image, which is an ambition for many dispositional essentialists (e.g. Ellis 2001). Why not just accept, since you have accepted the reality of properties like momentum, that it is momentum that is responsible for both resistance to change (inertia) and ability to exert forces on other balls? Why insist that in addition to having momentum an object has a particular power for every kind of visible consequence of interactions between objects with momentum (power to make other balls move, power to change one's own state of motion, power to make a hollow in a pillow, power to pull trains, power to break windows), especially when you can't avoid appealing to momentum when explaining all these various consequences? Do we want to say that the ball broke the window because it had the power to break windows? These conceptual quirks are a worry in addition to the main problem, that powers-based accounts do not take into account the established fact that whenever any object whatsoever acts on any other object whatsoever, the latter always acts on the first in the same way to the same magnitude and at the same time.

The identity theory of powers identifies an object's power to change its own state of motion, as well as that of other objects, with momentum. Hence it does not add to the set of natural properties defined by the sciences an infinite set of powers or dispositions corresponding to every distinguishable kind of behaviour. However, with the notable exceptions of myself (Ingthorsson 2002, 2021) and John Heil (2012), proponents of the identity theory have not generally taken to describing interactions in any other way than in terms of mutual manifestations of reciprocal disposition partners, where passive and active powers jointly contribute to a change in the object with the passive powers. I outlined what I think is wrong with that view in §5.

It is worth mentioning that both C. B. Martin (1993) and Mumford and Anjum (2018) are sceptical about the distinctions between active/passive and agent/patient and so do not characterize mutual manifestations in those terms. But they do not base this scepticism on the fact that whenever any object whatsoever acts on any other object whatsoever, the latter always acts on the first in the same way to the same magnitude and at the same time. In conclusion, powers-based accounts are by and large incompatible with the reciprocity of interactions, although this is not true of one or two alternatives. However, those whose accounts are compatible with the reciprocity of interactions are compatible for completely different reasons than the ones I outlined in §5.

7.4 Reciprocal Action between Powerful Particulars

According to the powerful particulars view, colliding balls each have their own directed momentum p_1 and p_2. Momentum is at the same time the power to resist changes in the state of motion and to change the state of motion of other balls. On contact the balls exert an equal and oppositely directed force on each other, that mutual exertion of influence being the cause of a change in their respective momenta from p_1 to p_{1*} and p_2 to p_{2*}. To repeat: the cause is the interaction as a whole and the effect is the sum total of changes suffered by the interacting entities. This formula applies to any collision regardless of initial state of any ball.

Note that there is an unresolved issue about the exact understanding of the nature of force. I lean to the understanding that forces are not properties of the balls either prior to collision or during the collision. To be exact, I do not consider them properties in virtue of which the balls exert influence; forces do not really push or pull, objects do. I understand the terms 'force', in this particular case, to denote the magnitude of the influence that balls with momentum exercise on contact. Similar understanding can apply to the force objects exercise in virtue of various other powers—for example, those related to charge, spin, etc. In this sense the concept of force is an abstraction, not least in light of the fact that forces only arise in interactions, and the fact that interactions are not considered to be composed of two separate entities, the action and reaction. However, by saying that the concept is an abstraction I am not saying it does not relate to a real phenomenon. The real phenomenon is reciprocal action, the mutual exertion of influence between two entities, and the abstraction is the conceptualization of reciprocal action as having two sides because it affects two (or more) entities. My understanding of forces comes close to that of Johansson (1989: 167–8) and Massin (2009), in that they treat them not as properties of an object but as something real that essentially holds symmetrically between objects, but I am uneasy about understanding reciprocal action as a relation. If it is a relation, it is a very special relation since it is an efficient relation and so more like an activity or process. Note that Massin, like Bunge, takes mutual forces to be non-causal relations because he thinks of them as symmetrical and therefore as unable to be productive, and yet he thinks they are relata of production. My view also has affinity with Jessica Wilson's view that forces are aspects of the objects that exert them and therefore not something in addition to the objects and their properties, although I am uneasy with her description of the objects as non-causal entities (Wilson 2007). As I mentioned earlier, Newman has recently offered an analysis of the collision of billiard balls that coincides with mine (Newman 2022: 307–8).

8. Case Study II: Water Dissolving Salt

The scientific explanation of water dissolving salt appeals, first, to the properties of H_2O and $NaCl$ molecules. Molecules of H_2O are covalent dipole compounds (have negatively and positively charged poles). $NaCl$ is a non-polar ionic compound. When molecules of H_2O and $NaCl$ come into contact there will be attraction between H_2O and either the Na or Cl in $NaCl$, depending on the spatial orientation of H_2O (negative pole attracted to positive ion, positive pole to negative ion). The covalent bond between O and H in water is stronger than the ionic bond between Na and Cl, which is why the tug of war between the various compounds (Na and Cl also attracting each other) ends in the breaking of the ionic bond of NaCl but not the covalent bonds in H_2O. The dissolution continues as long as individual water molecules can interact with individual NaCl molecules. It is important to note that the mutual attractions between the parts of each compound are all considered to be reciprocal.

8.1 Transmission Accounts

Transmission accounts are bound to explain the dissolution of salt in water in terms of transmitted quantities, but I have been unable to find in the philosophical literature any attempt to do this. Transmission accounts typically chose examples from the domain of thermodynamics such as the heating or cooling of water (e.g. see Fair 1979; Kistler 1998). I myself find it difficult to see an explanation of water dissolving salt solely in terms of transmission of conserved quantities.

The standard explanation of why and how water dissolves salt—the one given above—appeals to electrical charges and electrostatic interaction. One can, however, come across thermodynamic accounts of solubility, some of which give the impression of *explaining* solubility rather than just *describing* the thermodynamic aspects of dissolution. They claim to explain what happens in terms of systems striving for equilibrium. However, in so far as they only appeal to least-energy principles or say that NaCl breaks because that requires less energy than H_2O breaking, they turn out to be half-explanations. The reason that the breaking of NaCl is the most energy-efficient outcome is because in water covalent bonds are stronger than ionic bonds, wherefore it takes more energy to break the covalent bonds. But the relative strength of those bonds is not decided by transmission of conserved quantities.

In the end it seems difficult to explain attractions of any kind *solely* in terms of transmission of conserved quantities or striving towards equilibrium. Indeed, as Tracy Lupher argues, it is even difficult to explain any kind of static interaction in terms of transmitted quantities (Lupher 2009). I think his result is really the same as the more general conclusion I have reached that transmission accounts fail for symmetric interactions, whether dynamic or static.

The criticism here is that even in the domains of physics and chemistry that operate with conserved quantities, transmission accounts can at best only be applied with some plausibility to asymmetric interactions, but then only as approximations.

8.2 Mechanistic Accounts

Mechanistic accounts can easily be applied to explain the dissolution of salt by water. All the interactions taking place are either between the parts of organized wholes, or between organized wholes that consequently morph into other kinds of organized whole where interactions continuously preserve the whole. My only complaint is that NMP does not really address the question of unidirectionality vs. reciprocity of the interactions, wherefore it appears NMP can treat interactions sometimes as reciprocal when a given scientific explanation clearly tells us so, and sometimes as unidirectional when that is suggested by the scientific account. That raises the question of whether the deciding factor in treating any given interaction as unidirectional or reciprocal hinges on whether the scientific explanation for that particular interaction is considering the whole phenomenon or is only concerned to explain one side of the transaction. This ambiguity derives, arguably, from the fact that its proponents have not been aware of the problems I have raised here and elsewhere, and so have not taken any measures to respond to them.

8.3 Powers-Based Accounts

Powers-based accounts explain water dissolving salt in two different ways, depending on whether they relate only to the manifest or also to the scientific image. According to the former approach, the power to dissolve salt is attributed to water as a body of matter, and the power to be dissolved is attributed to salt. Accordingly, when salt is put in water the two powers mutually manifest the dissolution of the salt. This kind of explanation is open to two objections: (i) that it presents causation as a phenomenon involving unidirectional influence of the kind the natural sciences say does not exist and (ii) that it simply does not even address the fact that the scientific explanation of the phenomenon ties the ability to dissolve not to water as a body of matter, but to the properties of individual H_2O and $NaCl$ molecules. That raises the worry that these accounts at best relate to the way we ordinarily think about water dissolving salt, but not to what really happens.

The second approach takes the scientific explanation as its starting point and says that covalent dipole H_2O molecules have the power to break the ionic bond in $NaCl$ molecules (Marmodoro 2017; Kuykendall forthcoming). When such particles interact, the power to break and the power to be broken manifest the breaking of $NaCl$. This explanation is also open to the objection that it assumes interactions are unidirectional and is therefore in conflict with the third law. However, it does *ground* the powers of water and salt on the physical properties of the molecules, and therefore cannot be said to ignore the scientific explanation. On the other hand, the account does not *identify* the powers with the physical properties recognized by the sciences. The assumption is that in addition to H_2O having one slightly positively charged end, and one slightly negatively charged end, which gives the molecule the ability to interact electrostatically with other charged molecules (to mutually attract or repel), it also has the more specific power to 'break NaCl'. As far as I know, chemistry does not postulate any such specific property in addition to the already mentioned physical properties, but it acknowledges that for NaCl to break is a known consequence of electrostatic interaction between H_2O and NaCl. Importantly, it is clear that the properties of H_2O alone are not enough to ground its power to break NaCl. It can only be said to have that power with respect to the specific properties of NaCl, notably that it is an ionic compound. Indeed, power-based accounts rarely explicitly address the forces of push and pull operating between the molecules, but only the consequence of the forces exerted—NaCl breaks—and then say that the strength of the covalent bond, and/or the strength of the ion–polar electrostatic interaction between water and salt, gives H_2O the power to break.

I should address explicitly that Anna Marmodoro (2017) and Davis Kuykendall (forthcoming) complain that my account of causation in terms of reciprocal action 'fails to capture the directionality of the causal process, which is underpinned by the different "actions" of salt on water and water on salt respectively' (Marmodoro 2017: §7). On the one hand, it is true that I say that molecules exert an equal and oppositely directed force on each other, and that the interaction is in that sense symmetrical. On the other hand, I say that this symmetry allows for very different outcomes for each interacting entity in so far as they are different from the outset. One and the same kind of influence can result in different types of change in each interacting entity. In other words, we can talk about reciprocal actions as always being symmetric in terms of magnitude of influence and yet distinguish between symmetric and asymmetric interactions when talking about (i) interactions between two similar entities that produce similar changes in both and (ii) interactions between two dissimilar entities that produce different changes in each. My account therefore perfectly well explains why NaCl but not H_2O is destroyed in the interaction, and apparently without having to appeal to the kind of directionality that Marmodoro and Kuykendall are talking about. Indeed, I argue we should not appeal to it if we are to arrive at an account that applies to the full range of interactions and in a manner consistent with the third law. My argument doesn't show that there is no way to consider one side of the transaction as 'graver', only that previous ways of motivating that conclusion conflict with established scientific facts.

I can in turn complain that Marmodoro and Kuykendall's accounts of water dissolving salt bear some signs of the kind of half-explanations I have mentioned before. To be fair, Kuykendall recognizes that the breaking of NaCl is not the only outcome. Another outcome is the production of two different kinds of formation, which significantly alter the properties of the resulting liquid. Several molecules of H_2O will surround each Na^+ and Cl^- ion because of strong ion–dipole interactions between them to form what is called a 'hydration shell'. We now really have a liquid that is no longer made up only of H_2O molecules connected by hydrogen bonds, but is a mixture of molecules, some of which are connected by hydrogen bonds but others by ion–dipole bonds. All of this, I argue, can be explained by reciprocal actions between the parts in the liquid. The point is that Kuykendall's explanation, detailed as it is, is not complete in relation to the whole phenomenon since it focuses almost entirely on the 'water dissolves salt' aspect, rather than 'water and salt merge to form saline'.

Mumford and Anjum indeed take the position that we shouldn't think in terms of water dissolving salt but instead that salt and water together produce saline, and similarly that sugar and water produce a sweet solution (2011: 123). However, while much more fully recognizing the reciprocity of interactions, it is not because they accept the kind of reciprocity I have been arguing for here, or even are aware of the problem stemming from the rejection of unidirectional action, and their account still retains a notion of 'agency'—that is, of something being the pivotal trigger to a change that upsets a state in equilibrium, say a match being struck in a place with oxygen and flammable material. Indeed, they don't think so-called 'countervailing' powers should be included in the notion of cause (2011: 33–4). So, while we often come to similar conclusions, our accounts are decidedly different.

8.4 Powerful Particulars Account

According to the powerful particulars view, the dissolution of salt by water can be explained fully in terms of reciprocal interactions between the component parts of water and salt. The mutual electrostatic attraction between H_2O molecules on the one hand, and the components of NaCl on the other, results in separation of NaCl into Na^+ and Cl^- ions and the formation of hydration shells around each ion. The result is saline. Not only does the powerful particulars view agree in this way with the scientific understanding of 'water dissolves salt' but also with the scientific understanding of the formation of H_2O and NaCl molecules respectively, as well as of the constellations of Na^+ and Cl^- ions surrounded by a hydration shell. It does so without postulating a distinct second order class of dispositions/powers. It is all down to a question of who is the winner in various tugs of war between different entities, tugs of war whose strength is determined ultimately by the different strengths of the fundamental interactions on which they are based. Indeed, reciprocal interaction is a viable candidate for being what can give rise to the kind of organized wholes that mechanistic philosophers call 'mechanisms'.

9. Conclusion

In sum, I have argued that transmission, mechanistic, and powers-based accounts do not offer a generalizable explanation of physical phenomena, mainly because they either assume that all influence is unidirectional or are open to it being both unidirectional and/or reciprocal. Transmission accounts appear unable to account for mutual attractive influence and so are unable to account for the creation of the kinds of organized whole that would fit the description of mechanisms. Indeed, mechanistic accounts also struggle to explain how previously unconnected entities could become parts of a mechanism, and so struggle to explain how interactions between such entities would count as causal. I have suggested that my account of causation in terms of reciprocal action would resolve that problem.

It bears mentioning that there is one salient feature of the world that I haven't worked out how, or whether, the powerful particulars account could explain, and which may perhaps be the main reason why Bunge, Marmodoro, and Kuykendall are persuaded that all interactions cannot really be reciprocal and that my account fails to account for some kind of causal directionality present in the world. I am talking of the kind of feedback loops that we find in biological systems—or, really, self-organizing structures whose parts act on each other in a manner that suggests linear progression in a certain direction. The Krebs cycle is a good example. Every particular interaction that takes place in the cycle appears to be reciprocal and yet the process as a whole is directed in such a way that it repeats the same pattern again and again. Indeed, this would apply to the explanation of the role of enzymes as catalysts, as Kuykendall (forthcoming) mentions, to maintain a directed process. An explanation of that kind of direction is needed, but I doubt it will come in the form of overthrowing the third law of motion and vindicate the active/passive or Agent/Patient distinctions. One possibility is that this kind of directionality can be grounded partly on the directionality of the relation between the successive states of any organized whole of reciprocally acting parts, A—the one-sided existential dependence between producer and product—but will have to be constrained somehow by the structure of a larger organized whole, B, of which A is a part. To address the viability of that idea is work for the future.

Acknowledgements

I am very grateful to Yafeng Shan, Ingvar Johansson, and an anonymous referee, whose comments and suggestions decidedly helped to improve on the initial draft of this chapter.

References

Aristotle (1998). *Metaphysics*, H. Lawson-Tancred (trans.), London: Penguin.

Bechtel, W., and Abrahamsen, Adele (2005). 'Explanation: A Mechanistic Alternative', *Studies in History and Philosophy of the Biological and Biomedical Sciences* 36: 421–41.

Bogen, J. (2008). 'Causally Productive Activities', *Studies in History and Philosophy of Science* 39: 112–23.

Bunge, M. (1959). *Causality and Modern Science*, New York: Dover. Reprint 1979.

Craver, C. F. (2007). *Explaining the Brain: Mechanisms and the Mosaic Unity of Neuroscience*, Oxford: Clarendon Press.

Craver, C. F., and Darden, Lindley (2013). *In Search of Mechanisms: Discoveries across the Life Sciences*, Chicago: University of Chicago Press.

Darden, Lindley (2008). 'Thinking Again about Mechanisms', *Philosophy of Science* 75 (5): 958–69.

Dowe, P. (1992). 'Wesley Salmon's Process Theory of Causality and the Conserved Quantity Theory', *Philosophy of Science* 59: 195–216.

Dowe, P. (2009). 'Causal Process Theories', in Helen Beebee, C. Hitchcock, and P. Menzies (eds.), *The Oxford Handbook of Causation*, Oxford: Oxford University Press.

Ellis, B. (2001). *Scientific Essentialism*, Cambridge: Cambridge University Press.

Emmet, Dorothy (1985). *The Effectiveness of Causes*, Albany: SUNY Press.

Fair, D. (1979). 'Causation and the Flow of Energy', *Erkenntnis* 14: 219–50.

Galavotti, Maria Carla (2022). 'Wesley Salmon', in E. N. Zalta (ed.), *The Stanford Encyclopedia of Philosophy*, Stanford, CA: Stanford University, available online: https://plato.stanford.edu/archives/fall2022/entries/wesley-salmon/.

Glennan, S. S. (2009). 'Productivity, Relevance and Natural Selection', *Biology and Philosophy* 24: 325–39.

Glennan, S. S. (2017). *The New Mechanical Philosophy*, Oxford: Oxford University Press.

Heil, J. (2012). *The Universe as We Find It*, Oxford: Oxford University Press.

Hellingman, C. (1992). 'Newton's Third Law Revisited', *Physics Education* 27(2): 112–15.

Hertz, H. (1956). *The Principles of Mechanics*, New York: Dover.

Hume, D. (1748). *Enquiry Concerning Human Understanding*, Harvard Classics, Vol. 37, Boston, MA: P. F. Collier & Son, 1910.

Ingthorsson, R. D. (2002). 'Causal Production as Interaction', *Metaphysica* 3: 87–119.

Ingthorsson, R. D. (2013). 'Properties: Qualities, Powers, or Both?', *Dialectica* 67(1): 55–80.

Ingthorsson, R. D. (2016). *McTaggart's Paradox*, New York: Routledge.

Ingthorsson, R. D. (2021). *A Powerful Particulars View of Causation*, New York: Routledge. Open Access: https://doi.org/10.4324/9781003094241.

Jaeger, G. (2021). 'Exchange Forces in Particle Physics', *Foundations of Physics* 51: 13, https://doi.org/10.1007/s10701-021-00425-0.

Johansson, I. (1989). *Ontological Investigations: An Inquiry into the Categories of Nature, Man and Society*, New York: Routledge. Reprinted 2004 by Ontos Verlag. Open Access: https://www.degruyter.com/downloadpdf/title/303559.

Kant, I. (1787). *Critique of Pure Reason*, N. Kemp Smith (trans.), New York: St. Martin's Press, 1965.

Kistler, M. (1998). 'Reducing Causality to Transmission', *Erkenntnis* 48: 1–24.

Kitcher, P. (1989). 'Explanatory Unification and the Causal Structure of the World', in P. Kitcher and W. C. Salmon (eds.), *Scientific Explanation*, Minnesota Studies in the Philosophy of Science, Vol. 13. Minneapolis: University of Minnesota Press, pp. 410–506.

Kuykendall, D. (forthcoming). 'In Defense of the Agent and Patient Distinction: The Case from Molecular Biology and Chemistry', *British Journal for the Philosophy of Science*, https://www.journals.uchicago.edu/doi/abs/10.1086/715470.

Lupher, T. (2009). 'A Physical Critique of Physical Causation', *Synthese* 167: 67–80.

Mach, E. (1919). *The Science of Mechanics: A Critical and Historical Account of its Development*, T. McCormack (transl.), Chicago: The Open Court Publishing Company.

Machamer, P. (2004). 'Activities and Causation: The Metaphysics and Epistemology of Mechanisms', *International Studies in the Philosophy of Science* 18(1): 27–39.

Marmodoro, Anna (2017). 'Aristotelian Powers at Work: Reciprocity without Symmetry in Causation', in J. Jacobs (ed.), *Causal Powers*, Oxford: Oxford University Press, pp. 57–76.

Martin, C. B. (1993). 'Power for Realists', in K. Cambell, J. Bacon, and L. Reinhardt (eds.), *Ontology, Causality, and Mind: Essays on the Philosophy of D. M. Armstrong*, Cambridge: Cambridge University Press, pp. 175–83.

Martin, C. B. (1997). 'On the Need for Properties: The Road to Pythagoreanism and Back', *Synthese* 112: 193–231.

Massin, O. (2009). 'The Metaphysics of Forces', *Dialectica* 63(4): 555–89.

Maxwell, J.C. (1877). *Matter and Motion*, Dover: New York.

Molnar, G. (2003). *Powers: A Study in Metaphysics*, Oxford: Oxford University Press.

Mumford, S., and Anjum, Rani (2011). *Getting Causes from Powers*, Oxford: Oxford University Press.

Mumford, S., and Anjum, Rani (2018). 'Powers and Potentiality', in K. Engelhard and M. Quante (eds.), *Handbook of Potentiality*, Dordrecht: Springer, pp. 261–78.

Newman, A. (2022). 'A Causal Ontology of Objects, Causal Relations, and Various Kinds of Action', *Synthese* 200(4): article 308, https://doi.org/10.1007/s11229-022-03752-5.

Resnick, R., Halliday, D., and Krane, K. (2002). *Physics* (5th ed.), New York: John Wiley & Sons.

Russell, B. (1912). 'On the Notion of Cause', *Proceedings of the Aristotelian Society*, New Series 13: 1–26.

Salmon, W. (1980). 'Causality: Production and Propagation', in E. Sosa and M. Tooley (eds.), *Causation*, Oxford: Oxford University Press, 1993, pp. 154–71.

Salmon, W. (1984). *Scientific Explanation and the Causal Structure of the World*, Princeton, NJ: Princeton University Press.

Steinberg, M. S., Brown, D. E., and Clement, J. (1990). 'Genius Is Not Immune to Persistent Misconceptions', *International Journal of Science Education* 12: 265–73.

Taylor, R. (1973). *Action and Purpose*, New York: Humanities Press.

Wilson, Jessica (2007). 'Newtonian Forces', *British Journal for the Philosophy of Science* 58: 173–205.

7

The Metaphysics and Epistemology of Causal Production

The Prospects of Variation to Trace the Transmission of Information

Phyllis Illari and Federica Russo

1. Why Bother with Causal Production?

The sciences and everyday life are replete with cases in which we, epistemic agents, are interested in how causes *produce* effects. We want to know how SARS-COV-2 produces an inflammatory response, how particles interact at the subatomic level, how being late will make you miss your train, or how failing to properly plug in your headset will result in not hearing much of the videocall you are in.

It has already been noted in the literature that causal talk across the sciences and everyday life is very diverse; an argument made at least since Anscombe (1975), and repeated by a number of scholars in the philosophy of causality (Cartwright 2004; Weber 2007; Longworth 2010; Godfrey-Smith 2010). In fact, precisely because causal language and methods are so diverse, pluralistic strategies have been tried in many places. Just to confine the discussion to contemporary philosophy of causality, notable pluralists are Nancy Cartwright (2004), who famously pointed out that one word ('causation') in fact means many things, and Erik Weber (2007), who argued that different concepts of causality (probabilistic and process-based) are suited to different scientific contexts, and of course the whole group of scholars who, in the past two decades, developed pluralistic approaches to evidence for causal claims (see, e.g., Russo and Williamson 2007; Campaner 2011; LaCaze 2011; Clarke et al. 2013; Reiss 2015; Parkkinen et al. 2018; Pérez-González and Rocca 2021).

In previous work, we have also defended a qualified version of causal pluralism, the *causal mosaic* approach (Illari and Russo 2014). We think causality is not to be reduced to one philosophical question or one scientific problem. We have already presented our causal mosaic and will not rehearse our arguments in full here. This is just to signal that our account of information transmission as causal production, which is the object of this chapter, is part of a much broader, pluralistic,

Phyllis Illari and Federica Russo, *The Metaphysics and Epistemology of Causal Production: The Prospects of Variation to Trace the Transmission of Information* In: *Alternative Approaches to Causation: Beyond Difference-making and Mechanism.* Edited by: Yafeng Shan, Oxford University Press. © Oxford University Press 2024.
DOI: 10.1093/oso/9780192863485.003.0007

perspective on causality and not an attempt to provide yet another monistic account. We have also presented and defended information transmission in previous work as the most general account of causal production (Illari 2011b; Illari and Russo 2016a; 2016b; Vineis, Illari, and Russo 2017; Vineis and Russo 2018). In this chapter, we will not rehearse our argument, but provide some further qualifications of the account. In line with our pluralistic approach to causality, we hold that the account of causal production in terms of information transmission that we are developing needs to be *added* to the library of useful causal concepts. We also think we need a concept of production (information transmission) *and* a concept of difference-making (variation), which can usefully classify some of the useful accounts of causation, although not all of them. More specifically, in this chapter, we focus on causal production from a metaphysical perspective, and explore its corresponding epistemology. Note that we think the *concept* of variation is equally interesting, but here we focus on the *epistemological strategies* of variation, and their relation to information transmission as production.

Our distinctive approach here is practice-based on the one hand, and from an agent's perspective on the other hand (following Russo (2022)). In asking what causal production is, we are interested in a metaphysics that can be widely applicable across scientific domains. Importantly, though, this metaphysics is not a priori, but is always the product of an agent's perspective, and in this sense our approach aligns with perspectivism (Giere 2006; Massimi 2022), constructionism (Floridi 2011a; 2011b), and ontoepistemology (Barad 2007), and aligns with the lengthy discussion of Russo (2022). In Illari (2011b), Illari and Russo (2016b), and Russo (2022) we formulated a number of desiderata, which we adopt but also further elaborate as follows.

A concept of production should:

- [scientific domains:] make sense across sciences, including physics, social sciences, life sciences, and particularly for cases of causal relations *across* these levels;
- [levels:] help us understand causal relations across micro and macro causes (and vice versa) and across factors of different natures (sometimes called 'inhomogeneous variables'); and
- [technology:] be able to return a meaningful metaphysics for highly technologized contexts, in which there is arguably an important element of *construction* (so causal relations are not in any *simple* way 'out there').

The chapter is structured as follows. In section 2, we present information transmission as a thin and general metaphysics of production, with advantages and complementarity with respect to other production accounts. We explain a little further that our approach to metaphysics and ontology is in line with constructionism Floridi (2011b), and the ontoepistemology approach of Barad (2007), and Russo (2022). This means we need to understand how human epistemic agents *come to establish* that (or whether) there is transfer of information. In section 3, we will explore the epistemology of information transmission. In particular, we will distinguish between what we call 'information transmission' and 'variation' epistemological strategies. We will show that to know about information transmission requires both epistemological strategies that seem allied with our concept of production (such as mark transmission and process tracing) and epistemological strategies that seem allied with our concept of difference-making (variational strategies such as observational studies, studying variation across similar and different things). We will discuss how these strategies work in practice for four examples, highlighting the central role of epistemic agents. We will argue, however, that variational epistemic strategies are typically needed to establish even information transmission; indeed, most successful epistemologies use mixed strategies. In section 4, we will draw some more general conclusions from our examples. The chapter as a whole will show that we need a metaphysics of information transmission as production, but that the *epistemology* of production that accompanies it is rather complex, and does not reduce to information transmission. In general, good epistemology tends to use mixed strategies. Also, many of these epistemologies require a whole variety of instruments (from statistical software to lab equipment, or more simply our perceptual apparatus), but we place human epistemic agents in a central position to explain how these epistemological strategies work. In this chapter, for the first time, we begin to put two sides of our work together explicitly, the causal mosaic and the information transmission account of production, and develop the epistemology of production that is at work in tracing the transmission of information.

2. Information Transmission as a Thin General Metaphysics for Causal Production

2.1 Causal Mosaic

In previous work, we carried out a substantial review and systematization of the literature on causation in the past 60–70 years and showed that the project of finding *the one* concept of causation that fits *any* scientific context whatsoever is highly likely to fail (Illari and Russo 2014). We characterized causal pluralism as the view according to which causality cannot or should not be reduced to one notion or kind of thing only. Causal pluralism, however, is not an entirely new enterprise either. There exist a number of pluralistic approaches—for example, about types of causing, about inferences and evidence, about the very concept of causation, and about methods for causal inference (Illari and Russo 2014, ch. 23).

The pluralistic approach we present and favour is broader in character than other, existing pluralistic approaches, and can help systematize the literature in useful ways.

We name our approach 'causal mosaic' because we think of specific approaches, such as in terms of processes, dispositions, counterfactuals, or manipulation and invariance, as *tiles* that need to be put next to one another to form an image. The image we form is the mosaic of causal theory, which is a dynamic image. We particularly like the metaphor of the mosaic, but like any metaphor in science and philosophy, it should not be taken literally, but more heuristically. Thus, for instance, the mosaic helps us in conveying the idea that tiles next to each other contribute to make an image appear, but we do not think that tiles are mutually exclusive, and in fact they often overlap, or with some due modification they can be used in different parts of the mosaic. Thus, for instance, depending on specific purposes, we may need a mechanistic theory of causality or a causal theory of mechanisms (for a discussion, see Gillies (2018, ch. 4)). Once we have an inventory of available approaches—the tiles—the question is how to arrange them for particular purposes. One purpose is to account for a very specific question (e.g. how to causally explain the economic crisis of 1929), another is to give a highly general account of what causation is, and anything in between (see Illari and Russo 2014, ch. 20). What counts as a tile, what its borders and overlaps are, and how we arrange them, depend strongly on how we epistemic agents set the purpose.

Causal mosaic has two motivations. The first, philosophical, is that no single concept is likely to fit all scientific, technological or policy domains. In this sense, our approach is maximally liberal, in that it allows for various kinds of causing as well as various causal methods. This philosophical motivation is further substantiated by the observation that there isn't one single philosophical question of causation, but at least five:

(i) Metaphysics (or ontology): what is causality? What are causal relata?
(ii) Epistemology: what concepts guide causal reasoning or govern causal knowledge?
(iii) Methodology: what methods can be used to discover/explore/confirm causal relations?
(iv) Semantics: what is the meaning of 'cause'/'causality' in natural or scientific language?
(v) Use: what can we do (or not do) in the presence/absence of causal knowledge?

We follow Cartwright (2007) who argues for prioritizing use as a philosophical question. Here, 'use' does not merely refer to action, but encompasses a whole range of activities, epistemic, material, and scientific, in accord with Chang's 'epistemic practices' (Chang 2011; 2014).

The second motivation is scientific. Just as there isn't one philosophical question only, so the sciences deal with different types of causal problem. We identified at least five:

(i) Inference: does C cause E? To what extent?
(ii) Explanation: how or why does C cause or prevent E?
(iii) Prediction: what can we expect if C does (or does not) occur?
(iv) Control: what factors should we hold fixed to understand the relation between C and E? Or how should we modify C so that E accordingly changes?
(v) Reasoning: what considerations enter into establishing whether, how, or to what extent C causes E, and using that knowledge?

In our view, the causal mosaic helps in making sense of the vast intellectual enterprise about understanding causality, and about finding out about causes in the sciences. It recognizes that, for any given causal concept, we need to understand what it does help with, and what it does not help with. For example, take classic debates on the notion of mechanism. Mechanisms are arguably helpful in explanatory practices in, for example, biology or neuroscience, but metaphysical concepts such as capacities or dispositions might better help address ontological aspects of biological phenomena, perhaps as complementary to mechanisms.

Although our approach of causal mosaic has a strong pragmatist flavour, it does not license an 'anything goes' strategy. We think of the philosophical questions and scientific problems exactly as what should guide the choice of appropriate notions for particular contexts. Making the mosaic for a particular context is primarily an exercise in conceptual clarity, and we acknowledge that, in practice, philosophical questions about and scientific problems of causality are highly intertwined and interconnected. Concepts and notions of causality are not intrinsically good or bad, or better and worse than others, but they are more or less appropriate *for a given purpose*. Thus, for instance, if probabilistic theories are criticized for not proving good enough for causal explanation, the problem is not with probabilistic theories per se, but with what we set as a target for these theories: they aim to address problems of inference, rather than explanation, and so they are to be located primarily in the realm of epistemology and methodology.

In sum, the goal of the causal mosaic approach is to offer guidance to scientists and philosophers in selecting and using notions that are appropriate for the philosophical questions and scientific problems of causality they intend to address. It is part of the debate, we think, to discuss in the open which 'tiles' best fit the intended function and why, but such debate is premised on the idea that causal theory has to be inherently pluralistic, if it has to meet and account for the practices of the sciences and of policy.

2.2 The Question of Production: Information Transmission

We locate the question of production at the level of causal metaphysics: what *is* causation? What *are* causes? We are particularly interested in the metaphysical question, understood as a question of *linking*: what is it that *links* causes and effects? The way in which we aim to answer this question is *not* to give *the one* metaphysical concept, but to offer a very general concept that can help shed light on as many contexts as possible, in combination with other concepts. Recall that according to our mosaic pluralism, we seek to add information transmission to our library of causal concepts, not to replace existing concepts. In previous work, we highlighted the complementarity of information transmission with 'mechanisms', 'processes', and 'capacities-dispositions' (Illari 2011a; Illari and Russo 2016b). Here, we focus instead on its complementarity with 'variation', as is examined in Russo (2009).

Let us begin with clarifying the meaning and scope of causal production. Illari (2011b) distinguishes between giving an account of causation in its entirety as production, as, for example, Salmon–Dowe do, and giving an account of production within a pluralist perspective such as causal mosaic. Illari examines candidate accounts in the metaphysics of causality for an account of production—processes, mechanisms, and capacities/dispositions—and argues that while they are all useful, they are also insufficient. Illari sets the stage to provide an account of causal production, and of causal linking, in terms of information. This line of work has been further pursued in joint work (Illari and Russo 2016a; 2016b). Russo (2022) further expands on this, including the complementarity of mechanisms and dispositions to information transmission. It is worth noting that ours is not the only philosophical approach in terms of information. John Collier (1999; 2011) was perhaps the first in the causality literature to develop a full-blown account of causality in terms of information, holding that causation is the transfer of a particular quantity of information from one state to another. Others appeal to information in different ways. For example, James Ladyman and Don Ross (2007, 263) speak of 'information-carrying relations' that scientists study widely. Holly Andersen (2017) uses information-theoretic approaches to the notions of causal nexus and of patterns, and Billy Wheeler (2018) discusses the prospects of an information transmission account in the context of big data. Given this interest and attention to the notion of information, we hope to provide a very general framework for thinking about information, capturing a core of agreement among these approaches.

In previous publications, we presented 'information transmission' as a thin metaphysics to account for causal production. By the term 'thin metaphysics' we mean a minimal metaphysical commitment towards what causation is or what causes what, and this minimal commitment gives us maximal flexibility to cash out causal production across micro and macro factors, across factors of different natures, or in highly technologized contexts.

The most general idea of 'information transmission' comes from early accounts of mark transmission, notably as developed by Hans Reichenbach and Wesley C. Salmon. Specifically, the original idea of Reichenbach was to characterize causal processes in terms of worldlines, and in these worldlines we can track genidentical events (Russo 2022, chs. 11–12). Simply put, to establish the identity of some object, individual, or event, we need to establish the diachronic identity of some state of affairs that pertains to said object, individual, or event. When Salmon further developed this idea of mark transmission, he adopted a counterfactual formulation. A process is causal, following the idea of mark transmission, if, and only if, *were* we to mark it, that mark would be transmitted to later stages of the process. A stock example is the 'dented car'. Imagine we dent a car; when the car moves, the dent moves along. This mark— the dent—can be detected at later stages of the process. But this is possible because the process that is marked is causal. Instead, if we try to mark the *shadow* of the car, for instance trying to deform it somehow, we will see that the mark—the deformation—will not travel along. This is because the shadow is not a causal process, or, better said, we are not looking at the shadow in relevant causal interactions with other processes. The original account of Salmon was largely based on Reichenbach's, and both took the Special Theory of Relativity as a fundamental constraint on (physical) causation. But the main issue with the early formulation of the process account was related to its *counterfactual* formulation: a process is causal in case *were* a mark introduced, it would be transmitted at a later stage. We lack space to reconstruct this debate in detail (which we did in previous work), but the important element to retain is that, in very many cases, this counterfactual element, and the idea of introducing some kind of *material* mark, proves problematic. Put briefly, Salmon gave up his own mark-transmission account because he acknowledged that you cannot always mark physical processes without changing them, and he didn't want to save his account from this and other problems by making it *purely* counterfactual, saying that if we introduced a mark, then the process would transmit it, even in cases where that would never actually be possible. Instead, Salmon wanted his account to be in terms of what is actually being transmitted (Salmon 1994). Yet, we can retain the idea of mark transmission if we can generalize the notion of process beyond a physics formulation (including using physics quantities such as energy or momentum as hallmarks of causal interactions), and most importantly liberating mark transmission from the idea that marks have to be *introduced*.

This is where the concepts of information and information transmission prove helpful. We can think of information as a mark in a relevant sense, because information can and is transmitted by and in causal processes. Interestingly, this is hinted at by Salmon in his earlier work, but never further developed. In his 1994 paper, Salmon said (1994, 303):

> It has always been clear that a process is causal if it is capable of transmitting a mark, whether or not it is actually transmitting one. The fact that it has the capacity to transmit a mark is merely a symptom of the fact that it is actually transmitting something else. That other something I described as information, structure, and causal influence.   (Salmon 1984, 154–7)

However, unlike the formulation from philosophy of physics, information is not a mark we introduce, but a mark that is already there: we epistemic agents can describe processes in informational terms, and track whether information is transmitted or not. The agent's perspective we advocate here is key because, although our claim is that causal production *is* information transmission, it is important to bear in mind that any claim about *whether* information is transmitted or not is in fact the product of us epistemic agents engaging in numerous techno-scientific practices, from observation to manipulation. Differently put, any claim about causal production is the result of an ontology that depends on epistemology, and this is the gist of 'ontoepistemology' and of 'constructionism'.

This, however, does not answer yet the question of what this information is that is being transmitted. We primarily work with the concept of semantic information, which is a rather qualitative approach. In this way, we do not reduce information to mathematized and formalized accounts such as Shannon-Weaver's, and at the same time we do not exclude the use of such accounts in specific situations. Likewise, uses of and reasoning about information are common in many life sciences, particularly biology and the sciences of mind and brain. Although semantic information is clearly difficult to quantify, we take this to be a virtue of the approach, as it makes it a flexible and versatile way of expressing aspects of the world into contents that are semanticized, and this from an explicit agent's perspective. This means, to be more concrete, that we *can* use Shannon–Weaver information or biological information, but that we can also interpret marks such as the dented car as information. Or, as we shall try to show in the next section, information is what will help give an account of causal production in a variety of contexts.

3. Tracing Transmission: Why Production Needs Variation

3.1 The Intertwining of Causal Metaphysics and Causal Epistemology

Our approach is to understand causal production as transmission of information, in the most general terms. But it is useless to have a positive account of what causal production *is*, with no account of how we human epistemic agents *know* whether some effect is produced or not. In other words, a meaningful and useful metaphysical account has to be accompanied by an epistemology.

Our arguments about information transmission and variation begin from two more general stances. First, as we have said, we are pluralists about causation and, according to our causal mosaic, questions of metaphysics and of epistemology, while distinct, are also *connected*. Second, we think that any account of causal epistemology or causal metaphysics has to consider that these are at least in some sense 'products' of us human epistemic agents. To repeat, we adhere to broad principles of constructionism (Floridi 2011b) and of ontoepistemology (Barad 2007); there is no 'view-from-nowhere' and no ideal rational agents, but real epistemic agents that make inferences, whether in ordinary or scientific contexts, which is a view clearly in line with perspectivism too (Massimi 2022).

Given our views, a natural assumption might be that finding out about production requires information transmission epistemological strategies, and finding out about difference-making requires variational epistemologies. But we will show here that this is not true, at least for production. Finding out about production can be done with information transmission strategies, but it often also requires variational strategies. Broadly, we need to trace links and marks, and compare similarities and differences across cases. However, as we will get to later on, typically we need both strategies. This is consistent with the evidential pluralism about medicine that we have developed in other work (Russo and Williamson 2007; Illari 2011a; Gillies 2011; Clarke et al. 2013; 2014; Parkkinen et al. 2018). Evidential pluralism, briefly put, holds that to establish a causal claim, one typically needs evidence of difference-making (for instance, of correlations) and evidence of production (for instance, of mechanisms). The thesis is primarily epistemological and methodological in character, focusing on what *evidence* is needed for us epistemic agents to establish claims about causality, and it has been discussed by us and others in contexts such as medicine (see references above) and recently also in social science (Moneta and Russo 2014; Shan and Williamson 2021). Our claim here, instead, specifically pertains to causal production, able to work across different scientific domains, addressing epistemological strategies for knowing about causal production. While the scope *is* different, evidential pluralism and our thesis about information variation do share a pluralistic approach to evidence.

From an ontoepistemological and constructionist perspective, we therefore ask: how do we know that some C has produced some E? How do we know that information has (or has not) been transmitted? We know by observing some phenomenon and checking whether something changes or has *changed* and how it has changed. This requires making explicit a number of 'parameters', from the background knowledge used, to the empirical data that are available, from the methods and techniques used for causal analysis, to the situatedness of the researcher(s)—which means carrying out observations at a specific *level of abstraction*.

In this section, our main job is to address this key question: how do we *know* about information transmission? We think that while the question can be formulated in simple terms, the answer turns out to be rather complex. We can distinguish broadly two kinds of strategy that epistemic agents use. There are the classic strategies like observational studies and randomized controlled trials, which mostly seem to match the idea of causality (difference-making) as variation, but there are also strategies that seem to match more the idea of causality (production) as information transmission, such as process tracing or literal marking. We will illustrate how these strategies are used by epistemic agents in various contexts, but we will show that, in general, strategies used in successful causal inferences are typically mixed and intertwined. *Both* kinds of strategy are used to infer causal production.

We will use four examples to illustrate this idea. First, we examine two simple stock examples: billiard balls colliding and watering a plant. These cases capture folk intuition about causal strategies, and also begin to illustrate more fundamental, scientific issues in simple terms. Specifically, these simple cases help us to introduce the variational and information transmission strategies that we think are at work in a vast number of techno-scientific practices. Then we move to two cases (or episodes in the sense of Chang (2011; 2014)) where we examine practices in molecular epidemiology and in astrophysics. We discuss how epistemic agents track and trace the transmission of information in the 'exposome approach'; this is frequently studied as a case of production, but the second case, the science of supernovae is not, so we offer novel arguments for this. We show how important it is to discuss these episodes to highlight how information and variation work together, and from the agent's perspective: we are interested in metaphysical questions, but metaphysics does not fall from the sky, it is still the product of agents, working in groups and subject to all sorts of social and epistemic dynamics.

3.2 The Transfer of Information in Billiard Balls Colliding

Billiard balls colliding is a stock causality example, descending at least from Hume. It is typically taken to be the paradigm of material forms of causation, physical processes intersecting, causal interaction, and exchange of conserved quantities. This seems to be as clear a case of information transmission as is available, especially if we cash out information transfer as exchange of some physics quantities. But how do epistemic agents know that information is transmitted? We re-examine this case from the agent's perspective.

One answer that seems obvious is that moving billiard balls is a process that we can simply watch. Tracing the process is as easy as it ever gets. We can also mark the process—for example, the classic idea of chalk passing from the cue, to the first ball, and onto the second ball. Detecting the mark at the end of the process confirms that information has been transmitted. However, notice that we do also use variational strategies to find out some kinds of things, even in this extremely simple case.

First, we can know that processes of two balls transmit information because we measure relevant properties. We observe that some of these relevant properties *change* after interaction. This helps us to distinguish between real processes and what Salmon calls pseudo-processes. We can compare the billiard balls with intersecting shadows of aeroplanes on the ground: we see the shadows on the ground, we see they intersect, but no change is detected in the trajectory of either shadow after interaction. Comparing the change (billiard balls) to no change (shadows) studies variation.

Second, epistemic agents need to identify relevant properties, and variational strategies of examining and comparing repeated experiments and trials are often vital to help drill down and identify and establish for sure the relevant properties, and whether they can be generalized. For instance, relevant properties are the weight of balls, their form, including how hard they are, and from which direction they are hit—and not, for example, their colour.

All of these strategies are useful in understanding even such a simple causal process.

In this case, to repeat, the kind of information that is transmitted is exchange of some physics quantity. Our story, however, is that this simple answer needs to be complemented with a story about *how* an epistemic agent can establish *whether* information is in fact transferred or not.

3.3 Omitting to Water the Plant

Omitting to water the plant is another stock example of the causality literature. Failing to water the plant causes it to die. Omission is likely the cause, but omission is metaphysically controversial because something that is *not* cannot cause something else. Thus, according to a traditional materialist perspective, there cannot be any transmission of information, unlike in the collision of billiard balls. Notice that legal reasoning is replete with cases of omission, and this *is* causal reasoning. In the legal context, questions also arise about responsibility and accountability (see, e.g., Moore 2009), but we set aside this debate. Here we focus only on the question of production and whether omissions can be properly considered as cases of causal production, re-analysing from information transmission and the agent's perspective. This, we think, legitimizes talking about information transmission: not because we can univocally and directly point to material, physical transfers of information, but because an epistemic agent can analyse the case and reconstruct it in terms of the transmission of (semantic) information.

An epistemic agent can observe a plant and notice that it is in a pretty good state up to a certain point. Yet at some point the plant is dead. Upon receiving information that the plant has not been watered enough, this omission becomes *informative* about the next state of the plant. There is a *change* (i.e. variation) in the plant's life process, in which *omission* of watering explains death.

The epistemic agent can study two kinds of information transmission process here: first, the ongoing life of the plant, maintaining the plant in (living) homeostasis, and, second, the plant's constant intake of the essentials of life, including water. If one were to conduct experiments on information transmission processes in the plant, one way of marking an information transmission process would be to stop watering it. But that doesn't tell you much without variational strategies.

A variational strategy to infer information transmission would be to apply Mill's methods, and notice that, for instance, the only factor that changed was watering the plant, which then becomes informative about the production. An interesting feature of this case is that, while billiard balls colliding gives the impression that information transmission is ultimately reducible (and to be reduced) to *physical* transfer of some physical quantity, in this case it starts becoming clearer that whatever transmission *in the world* is, there is an ineliminable element of construction by the epistemic agent.

3.4 Biomarkers Research and Information Channels

In earlier publications, we analysed the case of molecular epidemiology, and more specifically the use made there of biomarkers in order to establish causal links (Illari and Russo 2016b; Russo and Vineis 2016; Vineis, Illari, and Russo 2017; Vineis and Russo 2018). Scientists in this field are interested in establishing links between exposure (e.g. pollution or certain chemical hazards) and disease (e.g. specific types of cancer, asthma, or allergies). The big challenge of molecular epidemiology is to connect macro and micro factors. How do we know that certain chemicals cause cancer? Scientists go searching for biomarkers of exposure, then biomarkers of early clinical changes, then biomarkers of disease onset.

Molecular epidemiology is an interesting area to look at because statistical methods and experimental equipment, together with novel theorizing about the notion of exposure, made it possible to reframe questions about exposure and health outcomes, which classic epidemiology had previously dealt with at a rather coarse-grained level, at a very fine-grained level. For example, exposure research uses proteomics, metabolomics, or genomics (sometimes called the 'omics technologies') to look at the fine-grained level of proteins, metabolites, and genes in the cell, as the body reacts to environmental exposure, and disease starts to develop. While this is fascinating work, even this level of detail does not make it easy to establish links as we have discussed, for instance, in the previous cases of the billiard balls and the unwatered plant. For one thing, the strategies to establish the linking between exposure and disease are still about establishing information transmission, but we can never establish *continuous* linking, of the kind we may directly observe in two billiard balls colliding. We can only access certain points in the complex interacting thing that is the human body, and in its interactions with the environment. Notice, however, that this does not mean that the link isn't continuous, simply that *we epistemic agents* can't trace *all* the points between exposure and disease. This is why an agent's perspective is needed: *we* reconstruct the process, and how information is transmitted all along. This is why our account is constructionist or ontoepistemological: information transmission *is* causal production, but as human epistemic agents we establish *whether* information has been transmitted through a number of epistemic and ontic strategies (and often with the aid of various instruments and technological equipment).

This work involves quite a few different strategies. A lot of experimental analysis of biospecimens is done, at various omics levels. But once the data about omics are generated, the analysis is statistical. Researchers conceptualize their methodology as 'meeting-in-the-middle' (Vineis and Chadeau-Hyam 2011), which is a statistical methodology that cross-references two types of correlation: correlations between exposure to a given hazard X and some biomarker (to be specified), and then correlations between that biomarker and some clinical conditions. The idea is to start to get a grip on the body reacting to environmental exposures as early as possible, and find which such biomarkers are linked to early disease onset. So in this way, we are trying to do something like tracing the moving billiard balls, but we can only access them at certain points, and using lots of technologies. The correlations found can then be understood as joint *variations* that help us track the linking. It is important to note, however, that these statistical analyses do not operate in a 'vacuum', but are based on lots of background knowledge, including knowledge of biochemical mechanisms that make it plausible to search for links at, say, the proteomic level rather than at the metabolomic level. It is in this sense that in Illari and Russo (2016b) we talk of mechanisms as *information channels* which impose constraints on possible and plausible linking, and so indicate where we should go to find the links. So this case shows how even research that is strongly dependent on variation strategies can still be seen as aiming to trace information.

Even more recent projects in exposure research add an important layer of complexity to the analysis, considering not only biochemical but also social mechanisms, broadly construed. The question is, for instance, how adverse childhood experiences are part of complex causal pathways that start quite early in life, and that become visible as clinical conditions much later in life. Here scientists search for links between exposure and disease that are not just biological, but also social and biosocial, and for this reason we need *socio*markers, next to biomarkers (Ghiara and Russo 2019).

Once again, our understanding of information transmission is a thin metaphysics because our metaphysical understanding of information is at once minimal and maximally liberal. We can achieve that by working with a notion of *semantic* information, which can amount to something quite material as exchange of physics quantities in the case of billiard balls and also as something more epistemically oriented as the *reconstruction* of a link by epistemic agents in the biomarkers case.

3.5 Exploring Information Transmission across Mechanisms and Processes in Supernovae

SN1987A was a particularly important supernova (McCray (1997) offers a quick, accessible introduction). It exploded some 168,000 years ago, and then became visible from Earth in the late 1980s. Two issues stand out. First, it extended our understanding of the cause, the particular 'mechanism' of Type IIa supernovae, specifically what carried away much of the energy of the collapse (see Prialnik (2010) for a textbook presentation, Suzuki (2008) for a review; see also Walker 1987; Lattimer 1988). Second, there was significant work immediately following SN1987A's discovery on the cause of its various peculiarities, including that it was the explosion of a blue supergiant, its light curve, and its peculiar 'squashed figure 8' appearance: there were two dimmer rings forming a 'squashed figure 8' surrounding the more usual bright centre with a single bright ring around it (Joss et al. 1988; Podsiadlowski and Joss 1989; Woosley and Chevalier 1989; Podsiadlowski et al. 1991).[1] What caused them?

These two issues both raise questions of causal production. The first question is a general case question of what produces the kinds of bright flares that we see, and can be distinguished by their light curves from other types of supernova. What is the *general* kind of continuous process that causes the things we detect, what the scientists call the 'mechanism' of Type IIa supernovae? In the second question, we are trying to figure out what happened in a single-case continuous physical process. So scientists detect the bright flare of SN1987A, also detect the neutrinos passing through Earth, and try to infer what happened to cause the bright flare, and in between that explosion and the neutrinos getting to Earth.

In addressing the first question, lots was already understood about supernovae. The size of the star's core is crucial, and SN1987A clearly had a core higher than the Chandrasekhar limit (about 1.4 solar masses), so would fall into Type II (although see Murdin (1993), for difficulties typecasting supernovae). This was theoretically understood to occur once nuclear burning in the star's core had turned it to iron, whereupon changing to further elements does not release energy, ceasing the nuclear burning that supports the star, and leading to its collapse (Prialnik 2010).

---

[1] Images are widely available. See, for example, Hubble's image: http://www.spacetelescope.org/images/potw1142a/, or https://en.wikipedia.org/wiki/SN_1987A.

So the beginning of this kind of process was quite well understood, while the end was even clearer. Stars exploding make them flare far brighter, and this is something that had been recorded for centuries before SN1987A. What was mysterious was the exact mechanism in between. Specifically, the collapsing core releases an enormous amount of energy that physical law tells us must go somewhere. Something must carry off the energy. A contemporary theoretical model suggested that a neutrino blast could do it. SN1987A was the first detection of neutrinos from such a source, reported by neutrino detectors at IMB and Kamiokande (so scientists were not able to *compare* to any other cases). Nevertheless, the causal process was filled in: SN1987A's core, larger than 1.4 solar masses, turned to iron, became no longer able to support itself against gravity, and collapsed, releasing an enormous neutrino blast, and creating the bright flare visible for several years (Sato, Shimizu, and Yamada 1995). This is the missing part of the mechanism of this type of supernova, in large part established on the basis of fewer than 10 neutrinos actually detected! In one sense this was relatively simple. Neutrinos were theoretically understood, known to exist, and had established detection methods. In theory they are created by the blast, and literally travel out in all directions including to Earth, and then they were indeed detected on Earth the expected time later (given their expected travel time and compared, for example, with visible light). In this case information transmission consists of the literal travelling of physical particles. And while we can only detect the information (i.e. the neutrinos) at one end of the transmission, that is all right in such cases. This reinforces the idea that information transmission does not require scientists to identify all points of the continuous process, but that instead we epistemic agents need to identify the relevant ones, just as in the case of exposure science, or even in the simple case of the plant not being watered, reconstructing the process. In another sense, this episode involves a very complex *epistemic activity of a large community* to, first, reconstruct the arrival of neutrinos on the basis of a tiny number of data blips created in massive underground water-filled structures and, second, converge on a postulated mechanism of Type IIa supernovae on the basis of a blast inferred from so few neutrinos.

Addressing the second question was important, because it was immediately obvious that something else must have happened with SN1987A. It did not merely show as a bright inner dot surrounded by a bright ring, as is usual, but also showed a fainter squashed figure eight, and it had other peculiarities too. What caused this? Work was done to modify standard models to understand what might be different about SN1987A. Stellar structure models of main sequence burning are already modified to get models of what will happen at the end of a star's life, yielding multiple possible mechanisms of supernovae. Yet SN1987A was already a non-standard Type IIa supernova. Simulations suggested that significant roiling in the star core prior to collapse could account for a violent asymmetric explosion that could lead to two 'cones' blasting far into space, the ends of which would appear from Earth as a 'squashed figure 8' (McCray 1997; Chevalier 1992). To be sure, this is a very simplified story, cutting out a lot of enduring controversy, but for our purposes it will do. So the process for SN1987A was thought to be as the newly agreed mechanism of Type II supernovae described above, but with significant roiling in the star core prior to collapse, which caused the peculiarity of the 'squashed figure 8' appearance.

Unlike exposure research, scientists theorize about parts and activities they literally think exist in relatively unproblematic ways, such as the star's core, roiling, and cones. Even neutrinos were in some ways well validated before 1987. Like the billiard balls, how those parts move and what happens to energy is a constant part of any inferences and understanding. Thus, the case of supernovae also shows how information transmission can be used across micro and macro levels of reality, crossing from something as large as a star's core passing the Chandrasekhar limit, to something as small as a neutrino.

However, just as in exposure research, the agent's perspective is key. Unlike in the billiard ball case, where an agent may be able to get qualitative, direct observation of the transmission of information (and also get quantitative measurements of such transmission), the cases of exposure science and of supernovae show that strategies to trace information transmission need to be complemented, or aided, by epistemological, variational strategies to trace the transmission of information. Even in a case where significant elements of information transmission consist in the literal travelling of physical particles vast distances, a large community of epistemic agents reconstructs information transmission, using variational strategies that led us to stellar structure models, mechanisms of supernovae, and the building of the simulations that explained the peculiarities. This all involved studying stars, building models that generate the data we see, comparing differences between different stars, and modifying models according to what theoretically *could* change, to see if we can match the differences we see, and gradually building a case for some changes—causes—over others.

This again shows how our ontoepistemological approach works. While the case of SN1987A shares some features of the simple billiard ball case, it is far from simple. Our understanding of information as at once minimal and maximally liberal works well, and here we see that our knowledge of even relatively simple transmission of physics quantities is really understood by epistemic agents reconstructing the link using significant technologies, as in the biomarkers case.

4. Discussion and Conclusion

We hold a pluralistic approach to causality that we dubbed 'causal mosaic'. Within this pluralistic framework, we focused here, following previous work, on questions of causal metaphysics, and more precisely of causal production. This is the question of what *links* causes and effects, and we have been defending 'information transmission' as the most general account of causal production, able to account for cases across levels (micro–macro), across inhomogeneous factors (biological–social), and in highly technologized contexts (admittedly, we haven't discussed this extensively here, but the interested reader is referred to Russo (2022, ch. 12)). The question of causal production, however, is not exhausted within the boundaries of metaphysics, because we share views that consider the perspective of agents as vital in the reconstruction of 'what there is', such as for instance the constructionist framework of Floridi, the ontoepistemo-logical framework of Barad, and the perspectival framework of Giere and of Massimi—accounts that are collectively discussed in Russo (2022).

In introducing an explicit agent's perspective, the metaphysical question of causal production as information transmission needs to be accompanied with an epistemological question about how we epistemic agents *know* whether informa-tion is transmitted. In this chapter, we illustrated with four examples how infor-mation transmission needs epistemological strategies, based on variation, that help us to establish the existence of information transmission. Differently put, we have sketched here the main lines of an epistemology of causality *of epistemic agents*, which in science are typically organized into epistemic communities. From these epistemic agents' perspectives we can establish facts about causal metaphys-ics. From the agent's perspective, we can distinguish two broad categories of strategies, although we have also shown that these are typically both used, and indeed are quite deeply intertwined. For the sake of clarity, we keep them separate, in order to highlight their main features.

Beginning with information transmission strategies, these can be understood quite generally as directly tracking the information, or the channel it runs through. This can be done by literally marking the process, like the billiard balls, or channels, such as dropping a dye in water and detecting it later down-stream, to show that chemicals could be transmitted that way. This is related to process tracing à la Steel (2008), which can involve simply watching change happen, as in the case of watching moving billiard balls, but can also involve watching stuff be maintained, like the living plant.

We showed that in exposure research conceptualizing linking as information transmission is helpful, but empirical access to that informational linking is very difficult, pushing scientists to primarily variational strategies. Note that back-ground mechanistic knowledge does a lot to make some kinds of information transmission channels more plausible than others (discussed more extensively in Illari and Russo (2016b)). SN1987A presents a different challenge. There is a single case process we are trying to trace, particularly at the time aiming to figure out a missing step. Scientists face the challenges of studying both very small and very large entities, which are also mind-bendingly far away. They cope with severe limits on what naturally travels close to Earth, as all our empirical information on supernovae must. However, scientists are still in a sense trying to access different points on that continuous process.

In sum, information transmission strategies are particularly good for ruling in and ruling out possibilities for causal linking, probing new unknown things, and studying single or rare cases.

Turning to variation (which has been deeply studied by Russo (2009)), these strategies can be understood very generally as comparing across similar and different things, tracing back to Mill's classic methods. Standard much-discussed strategies, of course, include randomized controlled trials and observational studies, or analysis of population data, but we have examined other kinds of work.

Even in the simplest physical processes, when we need to establish relevant properties, we often need variational strategies as in the billiard balls. Turning to the classic example of omitting to water the plant, strictly speaking we can mark it, as stopping watering it is a mark. But we really need variational strategies here to be sure what's going on, such as comparing other plants that are regularly watered to those given no water, or holding fixed details of who was in charge of watering the plant.

Variational strategies are very widely used in exposure research, going beyond the classic established study methodologies to 'patchwork' variations across the hypothesized linking from environmental exposure to disease onset, building a different evidential picture than previously existing work such as broader obser-vational studies. SN987A might be a single rare process, but empirical challenges mean scientists still need lots of modelling of that type of process (and comparison with empirical data for other stars) to get us going. Unsurprisingly, the need for similar variational strategies is very clear for building understanding of Type IIa supernovae in general. Note that Russo (2009) develops an account of variational strategies in great detail, particularly for social science practices. But the key point there is to demonstrate how a broad variational approach can capture the ration-ale for both interventionist and a range of observational methodologies in social science. In sum, variational strategies are particularly good for finding fine-grained relevant properties and for establishing links when we can't directly establish links, for which counterfactual reasoning (a special kind of variational strategy) is very helpful.

As we have emphasized, though, these strategies are deeply intertwined, and typically used together. While we have drawn out and shown different kinds of method, to distinguish the strategies, even in the simplified examples, multiple strategies are useful. Epistemic agents use strategies in concert.

To conclude, epistemic agents, organized in communities, use both information transmission and variational strategies to establish the existence of information transmission. This allows them to reconstruct continuous linking (causal produc-tion) when it is only empirically accessible at certain points, as is the case in both exposure research and the astrophysics of supernovae. Combinations of strategies yield much more than one alone—this is the key idea of the 'causal mosaic' (Illari and Russo 2014). The agent's perspective, notice, is crucial. It is important to point out that a constructionist or ontoepistemological perspective does not entail anti-realist stances; quite the contrary is the case. There is no causal metaphysics in the absolute, a priori, or detached from us, and the ontoepistemological perspective also demands that our causal metaphysics tells us something useful about causal epistemology. It is high time that we have an explicit account of causal production in which the very real role of epistemic agents is central, and this is the main message of our contribution.

References

Andersen, Holly. 2017. 'Pattens, Information, and Causation'. *The Journal of Philosophy* 114 (November): 592–622. https://doi.org/10.5840/jphil20171141142.

Anscombe, G. E. M. 1975. 'Causality and Determination'. In *Causation and Conditionals*, edited by E. Sosa, 63–81. Oxford: Oxford University Press.

Barad, Karen. 2007. *Meeting the Universe Halfway: Quantum Physics and the Entanglement of Matter and Meaning*. Durham, NC and London: Duke University Press.

Campaner, Raffaella. 2011. 'Understanding Mechanisms in the Health Sciences'. *Theoretical Medicine and Bioethics* 32: 5–17.

Cartwright, Nancy. 2004. 'Causation: One Word, Many Things'. *Philosophy of Science* 71: 805–19.

Cartwright, Nancy. 2007. *Hunting Causes and Using Them: Approaches in Philosophy and Economics*. Cambridge: Cambridge University Press.

Chang, Hasok. 2011. 'Beyond Case-Studies: History as Philosophy'. In *Integrating History and Philosophy of Science*, edited by Seymour Mauskopf and Tad Schmaltz, 109–24. Boston Studies in the Philosophy of Science, vol. 263. Dordrecht: Springer Netherlands. https://doi.org/10.1007/978-94-007-1745-9_8.

Chang, Hasok. 2014. 'Epistemic Activities and Systems of Practice: Units of Analysis in Philosophy of Science after the Practice Turn'. In *Science after the Practice Turn in the Philosophy, History, and the Social Studies of Science*, edited by Léna Soler, Sjoerd Zwart, Michael Lynch, and Vincent Israel-Jost, 67–79. Routledge Studies in the Philosophy of Science. New York and London: Routledge, Taylor & Francis Group.

Chevalier, Roger A. 1992. 'Supernova 1987A at Five Years of Age'. *Nature* 355 (6362): 691–6. https://doi.org/10.1038/355691a0.

Clarke, Brendan, Donald Gillies, Phyllis Illari, Federica Russo, and Jon Williamson. 2013. 'The Evidence that Evidence-Based Medicine Omits'. *Preventive Medicine* 57 (6): 745–7. https://doi.org/10.1016/j.ypmed.2012.10.020.

Clarke, Brendan, Donald Gillies, Phyllis Illari, Federica Russo, and Jon Williamson. 2014. 'Mechanisms and the Evidence Hierarchy'. *Topoi* online first. https://doi.org/10.1007/s11245-013-9220-9.

Collier, John. 1999. 'Causation Is the Transfer of Information'. In *Causation, Natural Laws, and Explanation*, edited by Howard Sankey, 215–63. Dordrecht: Kluwer.

Collier, John. 2011. 'Information, Causation and Computation'. In *Information and Computation: Essays on Scientific and Philosophical Understanding of Foundations of Information and Computation*, edited by Gordana Dodig Crnkovic and Mark Burgin, 89–105. Singapore: World Scientific.

Floridi, Luciano. 2011a. 'A Defence of Constructionism: Philosophy as Conceptual Engineering'. *Metaphilosophy* 42 (3): 282–304. https://doi.org/10.1111/j.1467-9973.2011.01693.x.

Floridi, Luciano. 2011b. *The Philosophy of Information*. Oxford: Oxford University Press.

Ghiara, Virginia, and Federica Russo. 2019. 'Reconstructing the Mixed Mechanisms of Health: The Role of Bio- and Sociomarkers'. *Longitudinal and Life Course Studies* 10 (1): 7–25. https://doi.org/10.1332/175795919X15468755933353.

Giere, Ronald. 2006. *Scientific Perspectivism*. Chicago: University of Chicago Press.

Gillies, Donald. 2011. 'The Russo-Williamson Thesis and the Question of Whether Smoking Causes Heart Disease'. In *Causality in the Sciences*, edited by Phyllis McKay Illari, Federica Russo, and Jon Williamson, 110–25. Oxford: Oxford University Press.

Gillies, Donald. 2018. *Causality, Probability, and Medicine*. New York: Routledge.

Godfrey-Smith, Peter. 2010. 'Causal Pluralism'. In *Oxford Handbook of Causation*, edited by Helen Beebee, Christopher Hitchcock, and Peter Menzies, 326–37. Oxford: Oxford University Press. http://www.petergodfreysmith.com/CausalPluralismPGS-07-Final.pdf.

Illari, Phyllis McKay. 2011a. 'Mechanistic Evidence: Disambiguating the Russo-Williamson Thesis'. *International Studies in the Philosophy of Science* 25 (2): 139–57.

Illari, Phyllis McKay. 2011b. 'Why Theories of Causality Need Production: An Information-Transmission Account'. *Philosophy and Technology* 24 (2): 95–114. https://doi.org/10.1007/s13347-010-0006-3.

Illari, Phyllis McKay, and Federica Russo. 2014. *Causality: Philosophical Theory Meets Scientific Practice*. Oxford: Oxford University Press.

Illari, Phyllis, and Federica Russo. 2016a. 'Causality and Information'. In *The Routledge Handbook of Philosophy of Information*, edited by Luciano Floridi, 235–48. New York and London: Routledge.

Illari, Phyllis, and Federica Russo. 2016b. 'Information Channels and Biomarkers of Disease'. *Topoi* 35 (1): 175–90. https://doi.org/10.1007/s11245-013-9228-1.

Joss, P. C., Ph. Podsiadlowski, J. J. L. Hsu, and S. Rappaport. 1988. 'Is Supernova 1987A a Stripped Asymptotic-Branch Giant in a Binary System?' *Nature* 331 (6153): 237–40. https://doi.org/10.1038/331237a0.

LaCaze, Adam. 2011. 'The Role of Basic Science in Evidence-Based Medicine'. *Biology and Philosophy* 26 (1): 81–98.

Ladyman, James, and Don Ross. 2007. *Every Thing Must Go*. Oxford: Oxford University Press.

Lattimer, James M. 1988. 'Supernova Theory and the Neutrinos from SN1987a'. *Nuclear Physics A* 478 (February): 199–217. https://doi.org/10.1016/0375-9474(88)90846-9.

Longworth, Francis. 2010. 'Cartwright's Causal Pluralism: A Critique and an Alternative'. *Analysis* 70 (2): 310–18. http://analysis.oxfordjournals.org/content/70/2/310.extract.

McCray, Richard. 1997. 'SN1987A Enters Its Second Decade'. *Nature* 386 (6624): 438–9. https://doi.org/10.1038/386438a0.

Massimi, Michela. 2022. *Perspectival Realism*. Oxford Studies in Philosophy of Science. New York: Oxford University Press.

Moneta, Alessio, and Federica Russo. 2014. 'Causal Models and Evidential Pluralism in Econometrics'. *Journal of Economic Methodology* 21 (1): 54–76.

Moore, Michael S. 2009. *Causation and Responsibility: An Essay in Law, Morals, and Metaphysics*. Oxford: Oxford University Press.

Murdin, Paul. 1993. 'Supernovae Can't Be Typecast'. *Nature* 363 (6431): 668–9. https://doi.org/10.1038/363668a0.

Parkkinen, V.-P., C. Wallmann, M. Wilde, B. Clarke, P. Illari, M. P. Kelly, C. Norell, F. Russo, B. Shaw, and J. Williamson. 2018. *Evaluating Evidence of Mechanisms in Medicine: Principles and Procedures*. New York: Springer Berlin Heidelberg.

Pérez-González, S., and Elena Rocca. 2021. 'Evidence of Biological Mechanisms and Health Predictions: An Insight into Clinical Reasoning'. Preprint. PhilSci Archive. http://philsci-archive.pitt.edu/id/eprint/19210.

Podsiadlowski, Ph., and P. C. Joss. 1989. 'An Alternative Binary Model for SN1987A'. *Nature* 338 (6214): 401–3. https://doi.org/10.1038/338401a0.

Podsiadlowski, Ph., A. C. Fabian, and I. R. Stevens. 1991. 'Origin of the Napoleon's Hat Nebula around SN1987A and Implications for the Progenitor'. *Nature* 354 (6348): 43–6. https://doi.org/10.1038/354043a0.

Prialnik, Dina. 2010. *An Introduction to the Theory of Stellar Structure and Evolution*. 2nd ed. Cambridge and New York: Cambridge University Press.

Reiss, Julian. 2015. *Causation, Evidence, and Inference*. London and New York: Routledge.

Russo, Federica. 2009. *Causality and Causal Modelling in the Social Sciences. Measuring Variations*. Methodos Series. New York: Springer.

Russo, Federica. 2022. *Techno-scientific Practices: An Informational Approach*. Lanham, MD: Rowman & Littlefield.

Russo, Federica, and Paolo Vineis. 2016. 'Opportunities and Challenges of Molecular Epidemiology'. In *Philosophy of Molecular Medicine*, edited by Giovanni Boniolo and Marco J. Nathan, ch. 12. New York and London: Routledge.

Russo, Federica, and Jon Williamson. 2007. 'Interpreting Causality in the Health Sciences'. *International Studies in Philosophy of Science* 21 (2): 157–70.

Salmon, W. C. 1984. *Scientific Explanation and the Causal Structure of the World.* Princeton: Princeton University Press.

Salmon, Wesley C. 1994. 'Causality without Counterfactuals'. *Philosophy of Science* 61: 297–312.

Sato, K., T. Shimizu, and S. Yamada. 1995. 'Explosion Mechanism of Collapse-Driven Supernovae'. *Nuclear Physics A* 588 (1): c345–56. https://doi.org/10.1016/0375-9474 (95)00160-3.

Shan, Yafeng, and Jon Williamson. 2021. 'Applying Evidential Pluralism to the Social Sciences'. *European Journal for Philosophy of Science* 11 (4): 96. https://doi.org/ 10.1007/s13194-021-00415-z.

Steel, Daniel. 2008. *Across the Boundaries: Extrapolation in Biology and Social Science.* Oxford: Oxford University Press.

Suzuki, A. 2008. 'The 20th Anniversary of SN1987A'. *Journal of Physics: Conference Series* 120 (7): 072001. https://doi.org/10.1088/1742-6596/120/7/072001.

Vineis, Paolo, and Marc Chadeau-Hyam. 2011. 'Integrating Biomarkers into Molecular Epidemiological Studies'. *Current Opinion in Oncology* 23 (1): 100–5.

Vineis, Paolo, and Federica Russo. 2018. 'Epigenetics and the Exposome: Environmental Exposure in Disease Etiology'. In *Oxford Research Encyclopedia of Environmental Science*, edited by Paolo Vineis and Federica Russo. Oxford: Oxford University Press. https://doi.org/10.1093/acrefore/9780199389414.013.325.

Vineis, Paolo, Phyllis Illari, and Federica Russo. 2017. 'Causality in Cancer Research: A Journey through Models in Molecular Epidemiology and Their Philosophical Interpretation'. *Emerging Themes in Epidemiology* 14 (1): 7. https://doi.org/10.1186/ s12982-017-0061-7.

Walker, Terry P. 1987. 'Making the Most of SN1987A'. *Nature* 330 (6149): 609–10. https://doi.org/10.1038/330609a0.

Weber, Erik. 2007. 'Conceptual Tools for Causal Analysis in the Social Sciences'. In *Causality and Probability in the Sciences*, edited by Federica Russo and Jon Williamson, 197–213. Rickmansworth, UK: College Publications.

Wheeler, Billy. 2018. 'How to Find Productive Causes in Big Data: An Information Transmission Account'. *Filozofia Nauki* 26 (4): 5–28. https://doi.org/10.14394/ filnau.2018.0021.

Woosley, S. E., and R. A. Chevalier. 1989. 'Was the Millisecond Pulsar in SN1987A Spun Up or Born Spinning Fast?' *Nature* 338 (6213): 321–2. https://doi.org/10.1038/ 338321a0.

8

Rebel With and Without a 'Cause'

A Theory of Causation

Julian Reiss

1. Introduction

Natural language contains a rich vocabulary for expressing causal relationships and processes other than *cause* and its cognates. I can *cause* the door to be closed, but I can also *shut* it, *have* it closed, or *force* it closed. While this feature of our language has been recognised in philosophy since at least Elizabeth Anscombe's Cambridge inaugural lecture (Anscombe 1971), its consequences are still under-appreciated in the discussion of causation. The main goal of this chapter is to make a contribution towards rectifying this neglect. I will show, in particular, that there are good reasons for this linguistic richness, and that statements expressing causal relationships or processes but not using the word *cause* cannot generally be reduced to expressions which do.

I will then proceed to argue that only one theory of causation, the inferentialist theory (Reiss 2015), can handle this feature of our language satisfactorily. The secondary goal of this chapter is thus to challenge proponents of alternative theories to make sense of the way causal relationships and processes are repre-sented in natural languages.

2. The Linguistics of Causatives

A note on methodology first. This chapter is primarily about the meaning of causal or, as I shall call them, causative claims.[1] Causative claims are expressions that contain a causative such as *cause*, *shut*, or *force* as a main verb. Establishing the precise meaning of causative claims is, however, by no means a trivial and uncontroversial endeavour. To give an example from the philosophical literature, most subjects in an experiment responded that the professor *caused* the problem when asked whether he or an administrative assistant caused the problem that

---

[1] I discuss the implications of my views about the semantics of causation for the metaphysics of causation in Reiss 2019. The epistemological implications have to wait for another occasion.

Julian Reiss, *Rebel With and Without a 'Cause': A Theory of Causation* In: *Alternative Approaches to Causation: Beyond Difference-making and Mechanism.* Edited by: Yafeng Shan, Oxford University Press.
© Oxford University Press 2024. DOI: 10.1093/oso/9780192863485.003.0008

there are no pens left after both the professor and the assistant had taken one but the assistant and not the professor had the right to do so (Knobe and Fraser 2008). According to this reading of the result of the experiment, the meaning of *cause* has an important normative component. However, according to an alternative interpretation of the results, subjects in fact do not answer a narrow causal question but a broader question about accountability, which has both causal and normative aspects (Samland and Waldmann 2016). This interpretation is entirely compatible with the absence of normative components in the meaning of *cause*. It is not a straightforward matter to determine who is right in this debate.

In writing this chapter I wanted to avoid two things. The first is to assume a normative theory of meaning and let that theory answer questions about the meaning of causative claims. Since one of my goals in this chapter is to argue in favour of a particular theory of meaning (for causative claims), this would simply amount to question begging. But I also wanted to avoid grounding my analysis in my own or experimental subjects' intuitions. My own intuitions I regard to be too unreliable for the job at hand, and I am not an experimental philosopher.

What I chose to do instead is to look at the literature on causatives in linguistics. In so doing, I found a number of facts about the meaning of causatives that appeared to be relatively uncontroversial and yet were incompatible with standard philosophical treatments of causation. It is these facts I am going to report in this section and on which my further analysis is based. Now, it may well be that a modicum of confirmation bias helped in the selection of facts. However, if anything, I was biased towards Anscombe's view of the topic when I began and was surprised to find that *cause* is not an abstraction from other causatives but in fact a causative that describes a specific kind of causal relationship and is thus on par with other causatives that are relatively abstract but by no means fully general (such as *make* or *force*).

Let us begin by revisiting Anscombe's view. She wrote (Anscombe 1971: 9):

> How does someone show that he has the concept *cause*? We may wish to say: only by having such a word in his vocabulary. If so, then the manifest possession of the concept presupposes the mastery of much else in language. I mean: the word 'cause' can be *added* to a language in which are already represented many causal concepts. A small selection: *scrape, push, wet, carry, eat, burn, knock over, keep off, squash, make* (e.g. noises, paper boats), *hurt*. But if we care to imagine languages in which no special causal concepts are represented, then no description of the use of a word in such languages will be able to present it as meaning *cause*.

Anscombe's concern in this work was to sever the tight link between causality and laws of nature; the idea that causes make their effects inevitable on account of the laws of nature. In the context in which the quote appears, Anscombe asks how we come to have the concept of cause. Her answer is that *cause* is an abstraction from more concrete causal concepts (such as *scrape* and *squash*), with many of which we have immediate perceptual acquaintance. As we shall see, Anscombe was mistaken about this. So let us see what linguists have to say about *cause* and related terms.

Linguists refer to what Anscombe calls 'causal concepts' as 'causatives'. They are frequently discussed in the context of a phenomenon called causative alternation (e.g. Levin and Rappaport Hovav 1994). Causative alternation involves verbs with transitive and intransitive uses, where the transitive use of a verb expresses a specific causal relation whereas the intransitive use expresses an outcome or process without highlighting any specific causal relation (Levin 1993: 26–7). For example:

A.   The shirt shrank. (*intransitive*)

B.   The cleaner shrank the shirt. (*transitive*)

This particular type of causative alternation between A and B is also called inchoative/causative alternation as, when used intransitively, the verb describes a change of state, but the phrase does not mention its cause or causer. When used transitively, a cause or causer is added, and the subject of the phrase using the intransitive verb becomes the object.

There are also induced-action alternations:

C.   The rats ran through the maze. (*intransitive*)

D.   The scientist ran the rats through the maze. (*transitive*)

and others that don't neatly fall into either of these categories:

E.   The baby burped. (*intransitive*)

F.   The father burped the baby. (*transitive*)

Many of the latter kind of alternation show a more limited range of objects in their transitive use than they show subjects in their intransitive use. For instance, one can burp a baby, but not its father, although both babies and fathers can burp (Levin 1993: 32).

There are three mechanisms that turn non-causatives into causatives (or vice versa): a lexical, a morphological, and an analytical mechanism. In a *lexical* causative the idea of causation is part of the semantics of the verb itself. This can be the case either if the verb is ambitransitive—that is, it can be used transitively and intransitively—as in the examples given above, or if a different word is used as in the pair die/kill:

G.   The cat died. (*intransitive*)

H.   Curiosity killed the cat. (*transitive*)

Morphological processes include the addition of prefixes or suffixes and changes in the verb stem to express cause. These don't really exist in English, but a few present-day English verb pairs, such as lie/lay or fall/fell, are remnants of a Germanic word-formation type which was productive in the derivation of causative from primary verbs by means of a specific suffix (García García 2012).

Analytical (also called periphrastic) causatives, finally, operate by adding a verb clause to the intransitive verb. *Cause* itself is the most obvious example:

I.   I squinted. (*intransitive*)

J.   The sun caused me to squint. (*transitive*)

There are many more verbs in English besides *cause* that can be used that way: *make* ('The sun made me squint.'), *have* ('She had him give her a massage.'), *get* ('She got him to give her a massage.'), *allow/let* ('She allowed him to/let him give her a massage.'), *force* ('The sun forced me to squint.') and several others. These constructions can add a causer to most verbs, whether intransitive or transitive:

K.   She made him push harder. (*intransitive verb in the subordinate clause*)

L.   She made him push the shopping cart harder. (*transitive verb in the subordinate clause*)

The grammar of causatives is not directly relevant for the project of this chapter. However, the differences in constructing a causative are associated with important differences in *meaning*. R. M. W. Dixon, for example, writes (Dixon 2000: 61; emphasis added):

> Quite a few languages have two or more causative constructions, involving either different formal mechanisms or different marking of the causee.... *There is always a semantic difference* and it may involve one or more of nine semantic parameters...

Dixon's nine semantic parameters are the following (Dixon 2000: 62):

1. State/action. Does a causative mechanism apply only to a verb describing a state, or also to a verb describing an action?
2. Transitivity. Does it apply only to intransitive verbs, or to both intransitive and simple transitive verbs, or to all types of verbs—intransitive, simple transitive, and also ditransitive?

3. Control. Is the causee lacking control of the activity (e.g. if inanimate, or a young child) or normally having control?
4. Volition. Does the causee do it willingly or unwillingly?
5. Affectedness. Is the causee only partially affected by the activity, or completely affected?
6. Directness. Does the causer act directly or indirectly?
7. Intention. Does the causer achieve the result accidentally or intentionally?
8. Naturalness. Does it happen fairly naturally (the causer just initiating a natural process) or is the result achieved only with effort (perhaps, with violence)?
9. Involvement. Is the causer also involved in the activity (in addition to the causee) or not involved?

For brevity let us focus on just two parameters here: volition and directness. Different degrees of causee participation in the causing can be expressed by using different main verbs in the analytical construction:

M.   She let him massage her.
N.   She made him massage her.

This works both for animate and inanimate causees:

O.   John let the vase fall.
P.   John made the vase fall.

Let implies the removal of an obstacle. But for the obstacle, the effect would obtain even in the absence of the causer (or the causer's action). *Make* implies causation against the causee's will or, if inanimate, its natural state. There is also, of course, a difference between (at least some) analytical constructions and more or less closely related lexical causatives:

Q.   She killed him.
R.   She let him die.

What does *cause* mean? The first thing to note is that, unlike what Anscombe suggests, *cause* is *not* the prototypical causative in English (Dixon 2000: 36–7):

> The convention in present-day linguistics is that a grammatical label should be based on a word of Romance origin—hence 'causative'. From this has arisen the misconception that *cause* is the prototypical causative verb in English. It is not; *make* is. *Cause* is a causative verb but it has a more specialized meaning (implying indirect causation) than *make* and it is much less common.

An example illustrates the difference between *cause* and the closest related lexical causative. The wife took a hammer and bludgeoned her husband to death. She *killed* her husband. Alternatively, she might have hired a contract killer. If the killer were successful, we would say that she had her husband killed. We *could* also say that she caused him to die. But we can't say that she killed him. *Kill* is not synonymous with *cause to die* because *cause* implies *indirect causation* whereas *kill* implies a more direct involvement of the causer. To give another example, when Sara turns the knob and pushes the door, we correctly say 'Sara opened the door'. When, however, she lifts up a window, a breeze enters the room, and the door opens, the lexical causative 'Sara opened the door' would be inappropriate while it would be perfectly acceptable to say 'Sara caused the door to open' (Wolff 2003: 2–3). Conversely, 'Your daughter upset a glass of milk' is a perfectly understandable complaint on the nursery nurse's part, whereas 'Your daughter caused the glass of milk to be upset' would be met with puzzlement because it would indicate that the daughter's action somehow set an unknown chain of causal relations in motion, at the end of which stood the spilled milk.[2] If the nursery nurse wanted to blame the daughter for spilling the milk, she would have to make the degree of the daughter's involvement clearer.

Make may be the prototypical causative verb in English, but in fact the word is not entirely general either. 'To make someone do something', for instance, means to employ force or authority over someone in order to compel them to do what they are asked, regardless of their personal inclinations (Nadathur and Lauer 2020). *Cause* certainly does not have *these* implications.

That different causative constructions allow different inferences is an entirely general fact about the semantics of causatives—whether in English or in other languages.[3] This is why, generally, expressions using one causative construction cannot be translated into an expression using a different causative construction. 'She killed him' can, very roughly, be translated as 'She caused him to die', but the direct involvement is missing. 'He made her clean the garage' can, very roughly, be translated as 'He caused her to clean the garage', but the exercise of force or authority against her will is missing. 'They let him play' can (perhaps), very roughly, be translated as 'They caused him to play', but the granting of a permission is missing.

[2] Philosophers may not cringe at an expression such as 'The cue ball's strike caused the second ball's motion.' But this is neither an ordinary nor a correct use of the term. Note that Hume did not use the verb *cause* in his description of the example: 'Here is a billiard ball lying on the table, and another ball moving towards it with rapidity. They strike; and the ball which was formerly at rest now acquires a motion' (Hume 1739/1960, Abstract).
[3] Jinghpaw, a language spoken in Burma, expresses causation by a prefix, sha-, and a suffix, -shangun. If X decapitated Y (i.e. is directly involved), the prefix is preferred; if X saw Y unconscious in water and didn't rescue him (or let him die) or if X ordered someone to decapitate Y (both cases of indirect involvement), the suffix is preferred. See Dixon 2000: 68.

3. The Linguistics of Causative and Theories of Causation

I am not aware of any discussion of these facts about the linguistics of causatives or even a recognition that these facts exist. They matter, however, for our understanding of the concept of causation. Discussing an example from Nancy Cartwright, Chris Hitchcock has argued that causatives do not pose a problem for the view that no word other than *cause* is needed to describe the causal content of expressions (Hitchcock 2007):

> It is certainly true that the claim 'the carburetor *feeds* gasoline and air to a car's engine' conveys more than that the carburetor *causes* gasoline and air to be present in the engine. The word 'feed' has subtle shades of meaning. In its paradigm usage, it applies to the provision of food for consumption by a human being (or other animal). The use of the word 'feed' in this context thus naturally invites a simile: air and gas are to a car's engine as food is to a human body. The air and gas are fuel for the engine, just as food is our fuel. This kind of nuance would be lost if the bare word 'cause' were used instead. But now a question arises: is the extra richness infused by the use of the word 'feed' instead of 'cause' itself causal in nature? That is, should we really say that 'feeding' is a thick causal concept that outstrips the bare notion of cause, or should we rather say that the meaning of the word 'feed' outstrips its purely causal implications? It is not clear to me that the causal content of the claim is not exhausted by the claim that the carburetor causes gasoline and air to be present in the carburetor [*sic*: car's engine], even if it manages to say more than this.

The claim 'the carburettor causes gasoline and air to be present in the car's engine' does not, however, 'exhaust' the 'causal content' of the claim 'the carburettor feeds gasoline and air to a car's engine'. Instead, it expresses *a different kind of causal relationship*. There are at least three differences.

First, while *cause* is more abstract than *scrape* or *squash* or indeed *feed*, like all analytical causatives it licenses specific inferences not licensed by other causative constructions. *Cause* initiates a causal chain of events, and is therefore indirect. By contrast, many lexical causatives express direct causation. One does not scrape a pan by initiating a Rube Goldberg machine-style causal sequence the outcome of which is a scraped pan. If there were such a sequence, the scraping of the pan would occur at the very end. In that case, someone who gets the machine running might well say 'I caused the pan to be scraped' but not that he scraped the pan.[4] Likewise, it would be a stretch to say about an absentee father whose sole

---

[4] Nadathur and Lauer 2020 argue that 'cause' implies causal necessity. Understood as a universal claim this is too strong in my view, but it is both true that if A causes B, many other conditions have to be met for B to come about and that A is often necessary in the circumstances for B.

involvement in childcare is the paying of alimony (without which the children would starve, say) that 'he feeds his children', though in a sense it is true that 'he causes them to be fed'.

Second, *cause* is a success term whereas many more concrete causatives are not. We cannot translate 'John pushed Jane' by 'John caused Jane to move', not even approximately, because push is not a success verb. Contrast:

John pushed Jane, but she wouldn't budge.

with:

*John caused Jane to move, but she wouldn't budge.

The second expression is non-sensical. *Feed* in the sense of *providing food* or, as in the given case, *supplying material to a machine* is not a success term. Just as one can feed a child without the child being fed (because of some obstruction, say), a carburettor can feed the engine without there being gas and air in the engine because of some malfunctioning.

Third, more controversially, *cause* implies counterfactual dependence but many other causatives don't. Here is an expression using *make* (Nadathur and Lauer 2020: 8; emphasis original):

> I usually go to soccer camp in the summer. Last year I was thinking about going to band camp instead, and I could not make up my mind. Then I broke my ankle, which settled things. I am so happy the injury made me skip soccer camp. I had the best summer ever!

Here the substitution of *cause* for *make* simply doesn't work (Nadathur and Lauer 2020: 9):

> I usually go to soccer camp in the summer. Last year I was thinking about going to band camp instead, and I could not make up my mind. Then I broke my ankle, which settled things. I am so happy the injury *caused* me to skip soccer camp. I had the best summer ever!

According to Nadathur and Lauer, the reason for this difference is that *cause* entails counterfactual dependence but *make* does not. In this case, the counterfactual 'Had I not been injured, I would have gone to soccer camp' is false because the speaker was unsure about it. The injury resolved the uncertainty, but without the injury no particular outcome could be expected.

This third point is more controversial because many philosophers maintain that cause does not imply counterfactual dependence because of cases of redundant causation. If Billy and Suzy throw rocks at a window, the window shatters due to Suzy's throw but the shattering would have occurred also in the absence of Suzy's throw because of Billy's, Suzy's throw caused the window to shatter but there is no counterfactual dependence, according to some philosophers (for a discussion, see Reiss 2015: ch. 1). Let me offer an alternative interpretation here. What should be unambiguously true is that Suzy's rock *shattered* the window. *Shatter*, unambiguously, does not imply counterfactual dependence. There is nothing in the word *shatter* that tells us that there isn't a backup cause. The reason that our intuition tends to support the claim 'Suzy's throw *caused* the window to shatter' is due to the fact that the causative claim that uses *shatter* is true and that we, mistakenly, hold that a more concrete causative claim implies some corresponding claim using *cause*. However, as argued in this section, concrete causative claims do not, generally, imply corresponding claims using *cause*. Thus, in this case, 'Suzy's rock shattered the window' is true while 'Suzy's throw caused the window to shatter' is not.

While this interpretation seems to provide an elegant solution to the problem of redundant causation at least in this case, I do not have the space here to provide a full defence of it, and I don't need it for the point I am making. What is important here is that the substitution of *cause* for *make* in the soccer camp example does not work. Causatives other than *cause* are therefore ineliminable from descriptions of causal relations, be it in ordinary language or in science.

4. A Problem for Truth-Conditional Theories of Causation

The existence and ineliminability of causatives (in addition to *cause*) poses a fairly obvious problem for standard theories of causation. Most theories of causation that are currently on offer aim to provide truth conditions for causal claims. That is, they aim to state necessary and sufficient conditions under which the causal claim is true.

Here are some examples of truth-conditional theories.

1.   The Regularity Theory (cf. Psillos 2002: 19):[5]

C causes E iff

(a) C is spatiotemporally contiguous to E;

(b) E succeeds C in time; and

(c) all events of type C (i.e. events that are like C) are regularly followed by (or are constantly conjoined with) events of type E (i.e. events like E).

---

[5] Psillos does not endorse the regularity account. I cite him because he provides a neat statement of Hume's account. I have slightly changed the notation for consistency.

2.   The Counterfactual Theory (Lewis 1973: 563):

C is a cause of E iff there exists a causal chain leading from C to E. A causal chain is a final sequence of (actual, particular) events F, G, H,... such that G depends causally on F, H on G, and so on throughout. C depends causally on E if and only if the family of propositions $O(E)$, $\neg O(E)$ depends counterfactually on the family $O(C)$, $\neg O(C)$.[6]

3.   The Probabilistic Theory (Suppes 1970: 28):

An event $C_{t'}$ is a direct cause of E_t iff $C_{t'}$ is a prima facie cause[7] of E_t and there is no t'' and no partition $\pi_{t''}$ such that for every $F_{t''}$ in $\pi_{t''}$

(a) $t' < t'' < t$;

(b) $P(C_{t'}\, F_{t''}) > 0$;

(c) $P(E_t | C_{t'}\, F_{t''}) = P(E_t | F_{t''})$.

4.   The Manipulability Theory (Woodward 2003: 59):

C is a (type-level) direct cause of E with respect to a variable set $\mathbf{Y}$ iff there is a possible intervention on C that will change the value of E or the probability distribution of E when one holds fixed at some value all other variables Z_i in $\mathbf{Y}$.

5.   The Mechanistic Theory (Glennan 1996: 64):

Two events C and E are causally connected iff there is a mechanism[8] connecting them.

6.   The Disjunctive Theory I (Hall 2004):

C is a cause of E iff C produces E or E depends on C.[9]

7.   The Disjunctive Theory II (Longworth 2010: 314):

C is a cause of E iff (manipulability & probability raising) or (locality & transference)[10] or (counterfactual dependence).

8.   The 'One Word—Many Things' Theory (cf. Cartwright 2004; the following is Longworth's 2010 interpretation):

C is a cause of E iff C φs E, where φ is a 'thick causal concept' (or causative) such as *compress, attract, discourage.*

---

[6] Lewis maintains that to any possible event E corresponds a proposition $O(E)$ that holds in all and only those possible worlds where E obtains. See Lewis 1973: 562.

[7] $C_{t'}$ is a *prima facie* cause of E_t iff (a) $t' < t$; (b) $P(C_{t'}) > 0$; (c) $P(E_t | C_{t'}) = P(E_t)$. See Suppes 1970: 12.

[8] A mechanism underlying a behaviour is a complex system which produces that behaviour by the interaction of a number of parts according to direct causal laws. See Glennan 1996: 52.

[9] Production is an intrinsic, local relation, possibly involving lawlike sufficiency; dependence is counterfactual dependence according to Lewis's account.

[10] Locality is essentially mechanistic connectedness as in Glennan's account. Transference is the transfer of a conserved quantity. See Dowe 2007.

9. The Epistemic Theory (Williamson 2014: 269):

C causes *E* iff there is an ideal causal epistemology which, when applied to an evidence base consisting of all fundamental matters of fact, yields that causal claim.

As can be seen, a variety of expressions can be found in truth-conditional theories on both sides of the 'iff'. In particular, we find both reductive (1–3) and non-reductive (4–8) theories as well as monist (1–5) and pluralist (6–8) theories. Some theories use the verb *cause* (1, 9), others the expression 'is a cause' or 'is a direct cause' (2–4, 6–8), and one theory 'is causally connected'.

If what I argued in the previous section is correct, *no* such theory can be a fully general account of causation. I take it that all the expressions using cause or a derivative are intended to be theories of the causal content of claims. That is, they are meant to apply to any scenario in which a causing is happening, independently of whether the scenario can correctly be described using the term *cause* or another causative.

However, as described in the two previous sections, different causative expressions have different implications. Here is another set of examples that, I hope, drives my point home (cf. Reiss 2019: 38):

S. The father burped his child.

T. The father caused his child to burp.

U. The father made his child burp.

V. The father got his child to burp.

W. The father let his child burp.

S expresses a direct involvement; T is indirect; U expresses intentionality and the exercise of force or authority on the father's part and some degree of (perhaps, unintentional) resistance on the child's part; V expresses successful encouragement; and W the removal of an obstacle or a permission.

Prima facie, we need different theories for different causatives because they are associated with different permissible inferences. Perhaps there is a common core meaning in virtue of which these (and so many other causative claims) are all 'causal' in nature. In the previous section we have seen, however, that whatever the meaning of *cause* is cannot be that common core. Just as *make* or *let* licenses specific inferences, so does *cause*. The meaning of *make* or *let* is not *cause* plus *x*.[11]

---

[11] Nadathur and Lauer 2020 argue that *cause* implies causal necessity (in the sense of counterfactual dependence) whereas *make* implies sufficiency. That the first part of the claim is controversial was argued in the previous section. I also have reservations about *make* expressing sufficiency. Whatever

This leaves the possibility that the common core is something other than the meaning of *cause*. I do not have a knockdown argument that there isn't such a common core. After all, it is not easy to prove a negative existential statement. But given the considerable differences in meaning between the different causative constructions, it is hard to see what this common core is supposed to be—in particular when some of the implications seem to contradict each other (e.g. direct involvement vs indirect causation; exercise of authority vs granting a permission).

Thus, no truth-conditional theory can be a fully general theory of causation. At best, it can be a theory of *cause* or *make* or *enable* or *burp* or …

5. Further Problems for Truth-Conditional Theories

In this section I want to discuss two further problems for truth-conditional theories of causation: metaphysical anarchy and polysemy/contextuality.

5.1 Metaphysical Anarchy

Truth-conditional theories have to make assumptions about the kind, nature, and number of the relata of causation—and of course about whether or not causation is to be considered a relation at all—on pain of succumbing to counterexamples.

Take Lewis' theory. As we have seen, he takes the number of relata to be two (C and E, cause and effect) and the kind as 'actual, particular event'. But he has to make assumptions about the precise nature of the events as well. My slamming of the door is counterfactually dependent on my shutting of the door but the former does not cause the latter. So Lewis demands that causal relations obtain between distinct actual events. More precisely (Lewis 2004 [2000]: 78):

> C and E must be distinct events—and distinct not only in the sense of non-identity but also in the sense of nonoverlap and nonimplication. It won't do to say that my speaking this sentence causes my speaking this sentence, or that my speaking the whole of it causes my speaking the first half of it, or vice versa; or that my speaking it causes my speaking it loudly, or vice versa.

the father may have done to his child, he needed the cooperation of the child's digestive tract in order to bring about burping. Very few factors if any are sufficient for their effects. Their discussion illuminates what they want to say though. *Make* expresses the idea that once the effect-making event occurs, the effect is as good as inevitable. *Cause*, by contrast, does not carry this connotation. When a cause obtains, a preventer can still intervene because the cause is indirect. But the maker is that last factor that makes virtually certain that the effect will occur.

This isn't everything. Events have to be individuated in just the right way in order for the theory to have a chance to work. The prince received a letter from his duke moments before he was assassinated. Reading the letter made a difference to his brain state, and so the exact death he died was different from the death he would have died in its absence. Alternatively, reading the letter delayed his receiving the visitor—who would eventually assassinate him—by a few moments. Again, receiving the letter made a difference to the precise death the prince died. And so, the prince's particular death is counterfactually dependent on the duke's writing the letter. But clearly, the duke's writing a letter to the prince did not cause the prince's death. Lewis therefore requires events to be relatively coarse grained in order to avoid a counterexample such as this: the actual event of the prince's death and the possible event in which the prince dies without having received the letter are the same event. But events cannot be *too* coarse grained: eventually, the prince would have died anyway. And so that possible death must differ from the violent death he actually died (otherwise it would not be counterfactually dependent on the assassin's action). The degree of granularity has to be just right.

Every truth-conditional theory makes assumptions that constrain the kinds of thing that can be causes and effects. Causatives pose a problem for such assumptions because they allow the representation of an enormous variety of kinds of thing that can be causes and effects and the ways in which they are related. My proof is of the inductive type, but a few examples concerning major truth-conditional theories will suffice to support my claim that any general assumption about the metaphysics of causing will face counterexamples.

In the first example a causative represents a causal process such that the cause logically entails the effect, in violation of Lewis's principle of non-implication. The example is taken from a seminal paper on tumour angiogenesis (Folkman 1974: 2109; emphasis added):

> There is increasing evidence that tumor cells communicate with normal host cells. [The nerve growth factor] is an example. Certain mouse sarcomas *secrete* a factor that stimulates growth in neighboring sensory and sympathetic nerve cells. However, the secretion of NGF is limited to a few tumor types.

Secrete means *form and give off a secretion*. Nothing can secrete without producing a secretion. The cause in this case is a certain activity in which the tumour engages. Trying to mould the scenario into the Lewisian framework would make us regard (a particular) mouse sarcoma's secretion of NGF as cause-event, but this event includes the effect-event (the production/presence of the nerve growth factor). The point is that the tumour cannot engage in the activity without causing the effect. But then the effect is logically entailed by the cause. Interestingly, perhaps, the word 'secretion' can refer to either the process or the product.

The example of secretion is analogous to 'The chair gouged the floor'. One cannot gouge a floor without successfully producing the effect: the gouge on the floor. One can *drop* the chair or *push* it or attempt to gouge the floor and fail, but one cannot gouge the floor and fail. To the extent, then, that causal relations or processes in science are described by success verbs such as *secrete* and *gouge*, Lewis' principle of non-implication is violated.

The linguistic literature on causatives recognises this. Sentences using lexical causatives that express direct causation can often be construed as a single event. 'The chair gouged the floor' describes such a single event. There aren't two events here that are causally related in one way or another. Now, as we have seen, *cause* usually expresses indirect causation, and indirect causation cannot be viewed as a single event (Wolff 2003). That aspect of Lewis' theory, then, is correct for *cause* but not for lexical causatives.

Of course, in some cases one may be able to re-describe the causal scenario without a success verb: for example, 'His dropping the chair caused the gouge on the floor'. However, the content of this statement is quite different from the original statement. Among other things, both causer and causee are different.

Causative verbs that are success verbs pose a problem for all those truth-conditional theories that assume that C and E are independent events or event-types (e.g. 1–3, 5–7) because there is just one—causal—event, not two events that are 'causally related'. Not all causatives work that way.

Some mechanistic and all causal power theories assume a radically different ontology in which causing is not a relation between events or event-types. Machamer et al. (2000), for instance, propose a dualist ontology that recognises both substances and the activities in which they engage. Causal power theories differ in their details but tend to assume that causing X is due to properties having the causal power to X (e.g. Mumford and Anjum 2011). These theories do really well with secretion and gouging cases ('The chair has a causal power to gouge floors qua its mass and shape') but they run into opposite difficulties.

Science recognises causation by absences as in 'Lack of sunlight causes vitamin-D deficiency'. Some more examples (Caceres et al. 2000: 173, quoted from Schaffer 2004: 202):

Muscular dystrophy in the mdx mouse has been described as a mutation in a colony of C57Bl/10ScSn mice, which results in the absence of the 427kDa membrane-associated protein dystrophin. . . . A deletion on the human X-chromosome causes the absence of an analogous protein and leads to Duchenne muscular dystrophy (DMD). . . . The absence of dystrophin leads to the destabilization of [the transmembrane glycoprotein complex], yielding weaker muscle fibers that undergo progressive degeneration followed by massive necrosis. Ultimately, premature death of DMD patients occurs . . .

There is no mechanism from the absence of sunlight to the absence of vitamin-D; nor is there a causal power. Mechanism and causal power theorists therefore resort to the following ploy: they stipulate that *cause* means *is mechanistically connected to* or *exercises its causal power to* unless it's causation by absence; in that case, *cause* means *counterfactually entails* (Dowe 2007: ch. 6; Mumford and Anjum 2011: 143–8). Dowe, for instance, introduces a new predicate for cases of negative causation: *caused** and defines it counterfactually.

This is unsatisfactory. Ordinary (and therefore also scientific) language draws distinctions along different lines than metaphysics. As we have seen, there are different ways to express causation and these ways correlate with semantic differences. There is no semantic parameter 'presence vs absence', however. Causation by absence will normally be expressed by an analytical rather than a lexical causative but this is because it tends to be indirect and the degree of control tends to be low. The neighbour who forgot to water my plants *let* them *die* but did not *kill* (or *dehydrate* or *desiccate*) them. Lack of vitamin-C *causes* scurvy, and there's no analogous lexical causative, but the reason is that there is a long temporal gap between cause and effect, and so the causing is highly indirect. The metaphysical theories do not carve causal language at its joints and therefore fail to track its meaning.[12]

The upshot is that even a truth-conditional theory that had '*C* φs *E*' (where φ is a causative) on the left-hand side of the 'iff' cannot work because if the theory does not provide constraints on the *C*s and *E*s that are being causally related, there will be counterexamples. However, any general constraints on what the *C*s and *E*s can be will also lead to counterexamples. The only way out of this dilemma is to replace the placeholders *C* and *E* with the values they take in a given application and provide an account of a *causative statement* instead of *cause* or the causative in between the *C* and the *E*. There would be one account for 'The chair gouged the floor', one for 'She let him massage her', one for 'Smoking causes lung cancer' and so on. Importantly, the account given for 'The chair gouged the floor' may well be different from that given for another statement using *gouge*.

5.2 Polysemy and Contextuality

Things get worse. This is because many causatives are polysemous. That is, causatives can have more than one meaning. Importantly: only some meanings

---

[12] These examples also show that the 'One Word—Many Things' theory (8) doesn't work as a truth-conditional theory because in some cases *cause* can (roughly) be paraphrased by a causative statement but not always.

are causal; other meanings are non-causal. Consider the—sometimes—causative *determine*. Here are some examples:

(1) The captain determined the position at sea.
(2) Two points determine a straight line.
(3) A reviewing committee determined a plea's validity.
(4) The family tried to determine the best time to go on holiday.
(5) The president determines national policy.
(6) Endogenous developmental factors determine cell fate.

Determine is used as a causative only in (5) and (6). In some cases, it is not clear from the sentence alone whether a causal relation is expressed. Let me give an example from economics (Robbins 1932: 81):

> The total quantity of money was opposed to the total quantity of goods and services exchanged against it; and its value per unit in any given period was held to be determined by the quantity multiplied by the velocity of circulation divided by the volume of trade—the celebrated equation of exchange $MV/T = P$.

In this quotation, *determine* is used functionally but it is at least conceivable that Robbins maintains a causal reading of the equation of exchange. These examples underline the point made in the previous section that a theory of causation must be a theory of *causative statements*. There can't be a causal theory of *determine* because *determine* can be used causally and non-causally. So we have to know what determines what in order to know whether the claim describes a causal relation.

The examples show more. In the last cases there is no unambiguous way to decide whether the claim is a causal claim on the basis of the claim alone. To decide this, we have to look to the *context* in which the claim was made. To learn whether *determine* is to be understood causally, we need to examine whether existing evidence in favour of the equation of exchange enables us to distinguish a causal from a functional reading. A theory of causation must be a *theory of causative claims in a context*.[13]

---

[13] Arguably, some existing truth-conditional theories are contextualist because they define causation relative to a variable set (e.g. the manipulability theory). The context then can be understood as picking out the right variables and values over which they range, and the theory used to identify causes within that context. There are three problems with this view. First, these accounts suffer from the problems discussed in Section 5.1 above. The manipulability theory, for instance, assumes that cause variables can always be intervened on, which is false. Second, any attempt to express a causative statement such as 'I opened the door' within a manipulability framework will change the meaning of the statement beyond recognition. Third, as I will argue below, such theories do not provide

6. The Inferentialist Theory of Causation

Let me, then, present an alternative to truth-conditional theories of causation. The main alternative to truth-conditional semantics (understood as a general theory of meaning) is inferential-role semantics (e.g. Brandom 1994, 2000). Without taking any stance on more general issues about meaning, I propose that the right way to think about the meaning of causative claims is in terms of the role they play in our inferential practices. More specifically, I propose that the meaning of a causative statement is given by the network of statements with which it is inferentially related.

A causative statement—such as 'Changes in the Bank of England's official rate move other short-term interest rates in the same direction', 'The tumour secretes a nerve growth factor', or, indeed, 'Smoking causes lung cancer'—is inferentially connected to all those statements from which it follows and which follow from it. Accordingly, the content of a causative statement is constituted by its relation to certain statements that describe the *evidence* for it, as well as statements that describe what I will refer to as the *cash value* of a causative statement. The evidence for a causative statement is given by observation statements, statements describing correlations, experimental results, facts that support hypotheses about mechanisms hypothetico-deductively, and so on. The cash value of a causative statement, in turn, is given by certain predictive and explanatory statements, by statements expressing effective strategies, and statements attributing praise and blame.

More precisely we can say that the content of a causative statement is given by its position in an *inferential system*. An inferential system for a causative statement is given by all those statements from which an epistemic community is entitled to infer the claim—the claim's *inferential base* (i.e. propositions describing its evidence)—and all those propositions the community is entitled to infer from it—the *inferential target* (i.e. its cash value). Regularities, correlations, interventions, mechanisms, and the like are not absent from the inferentialist account. But instead of directly providing the meaning of a causative statement, they enter its inferential base: that is, they *contribute to* the meaning of a causative statement but do not constitute it.

I use the expression 'entitled to infer' to indicate that the theory has a normative or rule-based component. It is clear, for instance, that a causative statement should not be inferred from the observation of a correlation if certain biases haven't been controlled for. Should this happen anyway, the content of the causative statement would not be constituted by the observed correlation alone. Rather, it would be given by all those propositions that would be needed to license

information about how the context picks out variables, assigns them values, etc. So even if this could be done in principle, an inferentialist account which provides such a mapping from context to semantics is preferable for the time being.

an inference to the causative statement (as well as those claims that would be licensed to infer from the causal claim) *in the given context.*

Contextualism in epistemology is now fashionable in analytic philosophy. When I speak of context, I don't mean *conversational* context, however, but *epistemic* context. The relevant epistemic context for a causative statement is given by (a) the nature and purpose of the causal inquiry; (b) domain-specific substantive information;[14] and (c) certain normative standards of the epistemic community.

Nature and purpose of the inquiry. A simple way to see how purpose enters the semantics of causative statements is the following. Suppose the aim of the inquiry is to establish why the match didn't light. Perhaps the match didn't light because it was damp. To explain an outcome, citing a necessary causal condition that makes a difference to the outcome is often enough. If the goal is policy, by contrast, a sufficient condition is needed. We have to know all the factors needed to bring about the outcome. A causal inquiry aiming at explanation then will seek to identify a necessary condition; an inquiry aiming at policy, a sufficient condition. These different purposes entail different evidential requirements to establish the causative statement.

Substantive information. Causal reasoning proceeds by eliminating alternatives. Say, our evidence for a statement '*C* causes *E*' is a study that finds a correlation between *C* and *E*. It is well known that a correlation between two variables has three causal explanations (*C* causes *E*, *E* causes *C*, some set of third factors *Z* cause *C* and *E*) and numerous non-causal explanations (e.g. selection bias, mismeasurement, Berkson's paradox, etc.). Domain-specific substantive information will tell inquirers which of these are relevant for a given case and need to be ruled out before the causal inference can be made. Domain-specific substantive information will also tell inquirers what observations they can expect to make under the supposition of the truth of a causative statement.

Normative standards. Causal inquiry proceeds against the backdrop of ethical, methodological, and conceptual commitments. To give just one example each, any piece of evidence can, in principle, be probed infinitely deeply. A study reports that *C* and *E* are correlated, but the reported correlation might in fact be due to a coding error in a spreadsheet. Probing, a researcher might re-run the calculation using her own application and find the same result. But the error might in fact be due to the data themselves being faulty. So she goes back to the statistical office where they came from to see if they have done their job properly. Even if they

---

[14] Aka background knowledge. I prefer 'substantive information' to 'background knowledge' because substantive information plays too important a role in causal inquiries for the label 'background' to be appropriate. It's really 99% existing information, 1% novel observations. 'Information' is preferable to 'knowledge' because I don't assume it to be true or even to be believed by inquirers. Truth is inaccessible and thus too high a standard. And inquirers must accept it for the purpose of the inquiry but not necessarily believe it.

have, there might be problems further down the line—for instance, because the people surveyed by the statistical office might have lied. There is no natural stopping point. One will stop once the cost of further probing exceeds the benefit of increased certainty of the result. But these are judgements that require ethical reflection. Any statistical analysis will have to take a stance on methodological questions such as whether to use frequentist, Bayesian, or likelihoodist principles. Finally, any causal inquiry proceeds against conceptual norms. Jacob Henle, a nineteenth-century German physician, made the following remarks about causes in medicine (Henle 1844: 25; quoted from Carter 2003: 24):

> Only in medicine are there causes that have hundreds of consequences or that can, on arbitrary occasions, remain entirely without effect. Only in medicine can the same effect flow from the most varied possible sources.... This is just as scientific as if a physicist were to teach that bodies fall because boards or beams are removed, because ropes or cables break, or because of openings, and so forth.

Henle was a proponent of the germ theory of disease according to which all diseases had a necessary universal cause. That theory enjoyed enormous successes during the nineteenth century but it ran into difficulties near the end of the century and was eventually abandoned altogether. The irony is that especially cancer causation is now back to a pre-germ theory situation as regards causes. In cancer causation, there are causes (say, smoking) that have hundreds of consequences (lung cancer, cardiovascular disease, impotence, poor vision, early menopause and tons of others) that can, on arbitrary occasions, remain entirely without effect; and the same effect (say, lung cancer) can flow from the most varied possible sources (say, smoking, asbestos, radon inhalation, genetics). One of the reasons for this situation is the way in which cancers are classified. Specific mutations are often necessary conditions for very specific kinds of cancer. We do not call the mutation 'the cause' of the cancer because it is harder to prevent and treat. Hence we focus on environmental and other factors. But what is useful for prevention and treatment changes, of course, with our knowledge and technological abilities. It is well possible that at some point in the future, cancer causation will follow a 'necessary universal cause' paradigm.

To examine the meaning of a causative statement, we have to identify its inferential system. To identify its inferential system, we have to identify its epistemic context. There is no precise mapping from context to content, but the inferential networks are fairly well understood. I have my own take on the topic (Reiss 2015), and recent work by philosophers as well as social, natural, and computer scientists on causal inference[15] and on using causative statements for

---

[15] To mention a just few, see Pearl et al. 2016 for a Bayesian approach; Mayo 1996 for a frequentist approach; Pawson 2006 for a critical realist approach; Illari and Russo 2014 for a survey.

explanations, policy, and attributing responsibility[16] indicates that maintaining that the meaning of causative claims is given by their inferential network is not only not empty but highly informative.

7. Inferentialism and Causatives

The inferentialist theory proposed here has none of the problems that beset truth-conditional theories. Recall that the problems discussed above had to do with (1) the lack of inter-translatability of causative statements; (2) the metaphysical anarchy represented by causative statements; and (3) the polysemy and context-uality of causative statements. Within an inferentialist framework, these issues can be addressed straightforwardly.

Lack of inter-translatability. The inferentialist theory doesn't make any distinction between causative statements that use *cause* and others that don't use *cause*. We can investigate the inferential system for any claim in language, scientific or otherwise. All statements will, to the extent that language users are justified in making them, in one way or another have been established and all statements will help to make inferences to other statements.

Different causative constructions embody different semantics, and a different semantics entails different inferential networks. Take *let* versus *make*. *Let*, as we have seen, requires a removal of an obstacle or the granting of a permission; *make*, a more active involvement of the causer such as force or authority and lack of consent on the part of the causee. To use 'She let him massage her' correctly, we need evidence of the permission; we also need evidence of his intention to massage her prior to the granting of the permission. 'She made him massage her' requires evidence of her exercising authority or force. Both evidence of her verbal or physical intervention and his mental state are needed in both cases; they differ between the two statements and thus they have a different meaning.

Establishing a statement that uses a lexical causative is no different. 'She bludgeoned him to death' requires evidence that (a) she hit him; (b) hitting him produced significant damage; (c) his death was due to the damage caused by the hitting. Specifically, there will have to be evidence to the effect that the nature and strength of the impact was able to cause his death and that he did not die because of some other factor, such as an existing brain aneurysm which accidentally happened to rupture at the same time as the bludgeoning. Different kinds of evidence will be admitted depending on the purpose—for instance, whether the statement is established in the context of a court case or for a historiography.

---

[16] Among many others, see Woodward 2003 for explanation; Cartwright 2007 for policy; and Moore 2009 for responsibility.

Metaphysical anarchy. Since the semantics of causation grounds in the reasoning practices in epistemic communities and not about the world, there is no need to make any assumptions about the metaphysics of causation. No metaphysical principles have to be assumed to be true of causal relations, certainly no general principles. What matters for the meaningfulness of causative statements is that there are rules for establishing them. And there are such rules whether the causer is an object, a state, a fact, an event, or a property, whether there are two events (a cause and an effect) or just one (a causing), or whether the result is brought about by a presence or an absence.

Polysemy. Since causal inferences are licensed, if at all, locally in the context of a given inquiry, the fact that causatives can mean different things in different contexts does not pose any problem. In fact, the inferentialist theory makes this feature expectable. Moreover, inferentialism has a good account of the alternate causal and non-causal use of some causatives. An inferential system can be given for any claim whether causal or not. A causal claim will, typically, have an inferential base composed of statements describing certain kinds of experiments and observational studies, statements describing the design of an experiment or observational study, and statements that describe facts helping to rule out biases such as selection and experimenter's biases and confounding. It will, typically, have an inferential target composed of statements describing explanations, predictions, effective strategies, and the attribution of responsibility. To what extent a given claim is causal can be determined by family resemblance with the inferential systems of paradigm causal claims. The more similar inferential base and target of the claim to a paradigm causal claim, the more causal it is. Causal and non-causal claims do not differ by kind, but by degree (see also Scriven 1966).

This last point also helps to address another problem. Why do we, for some purposes, group all causative verbs together despite significant differences in their meaning? Answer: because they are similar with respect to their inferential systems. Specifically, they all have observational statements concerning events, correlations among events, processes, mechanisms, and the like as well as statements ruling out alternative causes (and other explanations, say, of correlations) in their inferential base and explanatory, predictive and policy statements as well as statements attributing praise and blame in their inferential target. It is family resemblance with respect to inferential systems that allows us to classify the verbs as causatives.

8. Conclusion

Let me conclude by way of responding to an important criticism of the inferentialist account of causation (Williamson 2014). To get started, Williamson distinguishes a strong and a weak normative reading of the account, depending on how the inferential base and target are determined. According to the strong view, there is a prescriptive, external standard. Such an external standard might prescribe, for instance, that there always has to be evidence of difference making as well as evidence of a mechanism in the inferential base, as does the epistemic theory he defends (e.g. Williamson 2006). According to weak inferentialism, it is actually prevailing normative community standards that determine what a claim's inferential system is.

I defend weak inferentialism. There are no substantive, transcendent principles of scientific reasoning, whether about causation or any other matter. Williamson's own theory is a good example for the failure of an external standard. Causal claims can be established without evidence of a mechanism. That lithium is effective in the treatment of mania has been accepted for 80-odd years, but the mechanism is still not understood (Reiss 2012). Of course, we *could* argue that the scientific community was wrong to accept this claim. But it would be hard to make that case without already presupposing that evidence of a mechanism is required. The claim behaves exactly as if the use of the causal construction 'is effective' is felicitous. Lithium-based pharmaceuticals have been prescribed to millions of patients, and there is no news of such drugs being retracted. The claim is made (in one formulation or another) in countless publications—publications, which, again, have no history of mass retractions. We may be wrong, to be sure, but it is exceedingly implausible to say that the causal claim ought not have been accepted.

Williamson raises two problems for weak inferentialism. The first is excessive pluralism. Two claims that have different inferential systems mean, strictly speaking, something different. According to Williamson, this implies 'the word "cause" to mean different things in "inhaling tobacco smoke is a cause of cancer in mice" and "inhaling tobacco smoke is a cause of cancer in humans"', which is 'simply not plausible' (Williamson 2014: 266).

This criticism is, however, based on a misreading of the inferentialist account. The account does not answer the question: 'What does *cause* mean?' but is rather an account of the meaning of *causative statements in a context* because causatives can mean quite different things, depending on the sentences and contexts in which they are used. For example, there is no syntactical element that enables me to distinguish between 'The judge determines the sentence' and 'The quantity of money determines the interest rates' but one use is non-causal and the other causal. I could restrict the account to causal uses of verbs, but this would presuppose an understanding of causation, which the account aims to provide. Above I gave an example in which the very same sentence can be used causally and non-causally, depending on the (epistemic) context.

Having said that, the account is very radically pluralist. Unlike Williamson, I believe, however, that the degree of pluralism is palatable if we distinguish different levels of resolution. Strictly speaking, even the same causative claim can mean different things if the context changes—for example, because we have learned some new relevant facts. Suppose the mechanism through which lithium relieves mania becomes known. If so, this is likely to have both epistemic and practical ramifications. Many new lines of research will open up. And it is at least conceivable that more targeted drugs can be developed. It is *not* implausible to assume that these kinds of differences make a difference for the meaning of a claim.

Why do linguists maintain that *cause* means something different from *make*? Because the implications are different. *Cause* implies indirectness and, at least according to some linguists, necessity in the circumstances. Make implies the exercise of force or authority and a more direct involvement of the causer. In either case, evidence that these kinds of inferences are warranted must be possessed by the speaker in order to use the terms properly. So my theory is in line with linguists' practice, with the exception that the account is more holistic (for the given reasons).

While, strictly speaking, *cause* (and other causatives) can mean different things when used in different claims, and even the same causative claim can mean different things in different contexts, it is possible to examine similarities among different causative claims in their contexts. Causative claims that use 'cause' will be more similar to each other than to causative claims that use 'make' because the kinds of claim that make up their inferential bases and targets are more similar. Likewise, claims that employ the causal use of 'determine' will be more similar to each other than to claims that employ its non-causal use because inferential bases and targets are more similar. By the same token, but at a lower degree of resolution still, all claims using causative constructions share a high degree of family resemblance.

Williamson's second criticism is that weak inferentialism makes it impossible for a community to make *systematic* mistakes in causal reasoning. While it is possible for an individual speaker to ignore or misconstrue the community rules that govern the use of causative claims in their contexts, there is no sense in which the whole community can be wrong—because there is no external standard.

My response to this worry is that the rules, even if relative to a community of reasoners, are not arbitrary. By and large, causative claims are not established for purely cognitive reasons but because they are useful for making predictions, preparing interventions, providing explanations, assigning praise and blame, and so on (Reiss 2019). Some rules will be better at serving these purposes than others. Jacob Henle criticised contemporary medical practice for employing a concept of cause that, to his mind, was no longer useful. Today we criticise Henle for having defended a concept of cause that, from today's point of view, is flawed. But none of this implies that Henle and the subjects of his criticisms were using causal terms incorrectly. In other words, Williamson and I agree that reasoning practices can be systematically mistaken. He concludes from this that the practice employs causal language (more specifically, the verb *cause*) incorrectly. I conclude that causal terms are employed correctly, according to the norms of the community, but that the norms of the community are less than ideal for reaching its goals.

References

Anscombe, E. (1971). *Causality and Determination: An Inaugural Lecture.* Cambridge, Cambridge University Press.

Brandom, R. (1994). *Making it Explicit: Reasoning, Representing and Discursive Commitment.* Cambridge (MA), Harvard University Press.

Brandom, R. (2000). *Articulating Reasons: An Introduction to Inferentialism.* Cambridge (MA), Harvard University Press.

Caceres, S., C. Cuellar, J. C. Casar, J. Garrido, L. Schaefer, H. Kresse, and E. Brandan (2000). 'Synthesis of Proteoglycans Is Augmented in Dystrophic*Mdx*Mouse Skeletal Muscle.' *European Journal of Cell Biology* 79: 173–81.

Carter, K. C. (2003). *The Rise of Causal Concepts of Disease: Case Histories.* Aldershot, Ashgate.

Cartwright, N. (2004). 'Causation: One Word, Many Things.' *Philosophy of Science* 71 (PSA 2002): 805–19.

Cartwright, N. (2007). *Hunting Causes and Using Them.* Cambridge, Cambridge University Press.

Dixon, R. M. W. (2000). A Typology of Causatives: Form, Syntax and Meaning. *Changing Valency: Case Studies in Transitivity.* R. M. W. Dixon and A. Aikhenvald. Cambridge, Cambridge University Press: 30–83.

Dowe, P. (2007). *Physical Causation.* Oxford, Oxford University Press.

Folkman, J. (1974). 'Tumor Angiogenesis Factor.' *Cancer Research* 34: 2109–13.

García García, L. (2012). 'Morphological Causatives in Old English: The Quest for a Vanishing Formation.' *Transactions of the Philological Society* 110(1): 122–48.

Glennan, S. S. (1996). 'Mechanisms and the Nature of Causation.' *Erkenntnis* 44(1): 49–71.

Hall, N. (2004). Two Concepts of Causation. *Causation and Counterfactuals.* J. Collins, N. Hall, and L. Paul. Cambridge (MA), MIT Press: 225–76.

Henle, J. (1844). 'Medicinische Wissenschaft und Empirie.' *Zeitschrift für rationelle Medizin* 1: 1–35.

Hitchcock, C. (2007). How to Be a Causal Pluralist. *Thinking about Causes: From Greek Philosophy to Modern Physics.* P. Machamer and G. Wolters. Pittsburgh (PA), University of Pittsburgh Press: 200–21.

Hume, D. (1739/1960). *A Treatise of Human Nature.* Oxford, Clarendon Press.

Illari, P. and F. Russo (2014). *Causality: Philosophical Theory Meets Scientific Practice.* Oxford, Oxford University Press.

Knobe, J. and B. Fraser (2008). Causal Judgment and Moral Judgment: Two Experiments. *Moral Psychology: Cognitive Science of Morality: Intuition and Diversity* (vol. 2). W. Sinnott-Armstrong. Cambridge (MA), MIT Press: 441–8.

Levin, B. (1993). *English Verb Classes and Alternations: A Preliminary Investigation.* Chicago (IL), Chicago University Press.

Levin, B. and M. Rappaport Hovav (1994). 'A Preliminary Analysis of Causative Verbs in English.' *Lingua* **92**: 35–77.

Lewis, D. (1973). 'Causation.' *Journal of Philosophy* **70**(8): 556–67.

Lewis, D. (2004 [2000]). Causation as Influence. *Causation and Counterfactuals.* J. Collins, N. Hall, and L. A. Paul. Cambridge (MA), MIT Press: 75–106.

Longworth, F. (2010). 'Cartwright's Causal Pluralism: A Critique and an Alternative.' *Analysis* **70**(2): 310–18.

Machamer, P., L. Darden, and C. Craver (2000). 'Thinking about Mechanisms.' *Philosophy of Science* **67**: 1–25.

Mayo, D. (1996). *Error and the Growth of Experimental Knowledge.* Chicago, University of Chicago Press.

Moore, M. (2009). *Causation and Responsibility: An Essay in Law, Morals, and Metaphysics.* Oxford, Oxford University Press.

Mumford, S. and R. L. Anjum (2011). *Getting Causes from Powers.* Oxford, Oxford University Press.

Nadathur, P. and S. Lauer (2020). 'Causal Necessity, Causal Sufficiency, and the Implications of Causative Verbs.' *Glossa: A Journal of General Linguistics* **5**(1): 1–37.

Pawson, R. (2006). *Evidence-Based Policy: A Realist Perspective.* London and Thousand Oaks (CA), SAGE.

Pearl, J., M. Glymour, and N. Jewell (2016). *Causal Inference in Statistics: A Primer.* Chichester, Wiley.

Psillos, S. (2002). *Causation and Explanation.* Stocksfield, Acumen.

Reiss, J. (2012). Third Time's a Charm: Wittgensteinian Pluralisms and Causation. *Causality in the Sciences.* P. McKay Illari, F. Russo, and J. Williamson. Oxford, Oxford University Press: 907–27.

Reiss, J. (2015). *Causation, Evidence, and Inference.* New York, Routledge.

Reiss, J. (2019). 'Causal Explanation Is All There Is to Causation.' *Teorema* **38**(3): 25–52.

Robbins, L. (1932). *Essay on the Nature and Significance of Economic Science.* Toronto (ON), Macmillan.

Samland, J. and M. Waldmann (2016). 'How Prescriptive Norms Influence Causal Inferences.' *Cognition* **156**: 164–76.

Schaffer, J. (2004). Causes Need Not Be Physically Connected to Their Effects: The Case for Negative Causation. *Contemporary Debates in Philosophy of Science.* C. Hitchcock. Oxford, Blackwell: 197–216.

Scriven, M. (1966). Causes, Connections and Conditions in History. *Philosophical Analysis and History*. W. Dray. New York, Harper and Row: 238–64.

Suppes, P. (1970). *A Probabilistic Theory of Causality*. Amsterdam, North-Holland.

Williamson, J. (2006). 'Causal Pluralism versus Epistemic Causality.' *Philosophica* 77 (1): 69–96.

Williamson, J. (2014). 'How Can Causal Explanations Explain?' *Erkenntnis* 78: 257–75.

Wolff, P. (2003). 'Direct Causation in the Linguistic Coding and Individuation of Causal Events.' *Cognition* 88: 1–48.

Woodward, J. (2003). *Making Things Happen*. Oxford, Oxford University Press.

9

Causal Fictionalism

Antony Eagle

A few years ago now I argued that causation, while not a relation of fundamental physics, is nevertheless pragmatically indispensable (Eagle 2007a). A number of other philosophers made similar arguments around the same time (Price 2007; Menzies 2007), and there are many precedents in the literature. Here I want to revisit these arguments with the benefit of hindsight. I don't in essence disagree with what I said back then, but I think what I say here both significantly clarifies my earlier discussion and advances things to some degree.

In the first part of the present chapter, I start with the role of causal models in the human sciences (§1), and attempt to explain why it is not possible to straightforwardly ground such models in fundamental physics (§2). This suggests that further constraints, going beyond physics, are needed to legitimate such models (§3). These supplementary constraints could be reified, but that would seem to conflict with the completeness of physics (§4). A response is to emphasize the practical role of causal talk (§5), and I suggest that a fictionalist approach might be worth exploring. After clarifying fictionalism as a general approach (§6), in §§7–9 I try to carry out in some detail the project of clarifying what a fictionalist attitude to causation would involve.

1. Causation in the Human Sciences

Ramsey observed that

> from the situation when we are deliberating seems to arise the general difference of cause and effect. We are then engaged not on disinterested knowledge or classification ... but on tracing the different consequences of our possible actions.
>
> (Ramsey [1929] 1990, 158)

Cartwright (1979) makes the related point that causes are vital in drawing a distinction between effective and ineffective strategies for attaining our desired ends. When we are deciding which actions to perform, we ought generally to pursue actions which genuinely promote our goals. The standard decision theor-etic approach to deliberation says we ought to pursue those actions with the

Antony Eagle, *Causal Fictionalism* In: *Alternative Approaches to Causation: Beyond Difference-making and Mechanism*. Edited by: Yafeng Shan, Oxford University Press. © Oxford University Press 2024.
DOI: 10.1093/oso/9780192863485.003.0009

greatest expected value to us (Peterson 2017). Here 'expectation' is a probabilistic notion; Cartwright's central point is that the probabilities involved need to capture causal information (1979, 431).

Some strategies can be seen to be ineffective on purely statistical grounds. The probability of being susceptible to mind reading, conditional on wearing a tin foil hat, is very low. But that is due to the low unconditional probability of mind reading, which is probabilistically independent of tin-foil-hat-wearing. This is an ineffective strategy because it is probabilistically irrelevant. By contrast, there was a strong non-accidental association between the construction of airstrips and the delivery of clothing, medicine, and other technologically advanced goods through-out Melanesia during World War II. But the 'cargo cult' strategy of constructing mock airstrips didn't further the participants' goals, because the association was the symptom of a common cause.[1] Building airstrips was an ineffective strategy because it involved probabilistic dependence without causal relevance. An effect-ive strategy would have probabilistic dependence that was backed by causal relevance: for example, the strong association between the deployment of malaria nets and the decreased incidence of childhood malaria is backed by a plausible causal mechanism (Levitz et al. 2018).

The human sciences—the social sciences, health and medical sciences—are focused not on 'disinterested knowledge', but on providing a basis for action to improve people's lives (given some antecedent conception of wellbeing). It is not surprising, then, that even cursory glance at the literatures of the human sciences suggests they are replete with invocations of causation. Attributions of blame and responsibility, evaluations for the efficacy of interventions, and tests to discrim-inate causal from non-causal associations are basic parts of the standard concep-tual toolkit in economics (Hoover 2008), public policy (Cartwright and Hardie 2012), clinical medicine (Williamson 2019; Stovitz and Shrier 2019), and public health (Hill 1965; Lucas and McMichael 2005), among many others. The key results—and the principal reasons for our interest in these fields—are causal claims: claims that a certain drug, or public health initiative, will lower the incidence of negative outcomes from illness or disease (i.e., cause them to be less frequent); or that adoption of some policy will promote a desirable social or economic outcome. Across these fields we see a trans-disciplinary deployment of techniques of statistical causal inference to secure such results. These techniques aim to deliver exactly what Cartwright argued is needed to discriminate effective strategies: a way of identifying non-causal ('spurious') associations (Granger 1969; Simon 1977; Suppes 1970; Pearl 2000, 42–57).

However, these techniques are often strikingly empiricist, in the ancient sense (Dawes 2017, §3.3): theoretical mechanisms are discounted in favour of observed

---

[1] The real anthropological story, of course, is more complex (Lindstrom 1993).

associations among the values of random variables, preferably ascertained by a systematic review of randomized controlled trials (Howick et al. 2011). The evidence consists of observable values of random variables, a theoretically laden classification of 'raw experience' into usable data. These random variables partition ('coarse-grain') the space of possible outcomes into classes whose members all agree on the value of that variable. The variables whose values can be aggregated into statistical data cannot be arbitrary functions from outcomes—to be *data*, different outcomes must be observably distinct from one another, so that the value of the variable in a given state can be discerned. For these variables are to be set exogenously in the causal model, rather than having their values fixed endogenously within the model, and while purely theoretical variables could in principle play this role, in practice the values of exogenous variables are determined by statistical observation. These refined empirical data are then used to generate a causal model depicting a pattern of relations between the random variables (they may postulate unobservable hidden variables too). Generally, there are many models consistent with the data, and we prune the models by eliminating any that don't capture all and only the observed statistical dependencies between variables. This process can terminate in a unique model only in the case where there are no hidden variables. These models give us causal relations that accord, more or less, with key platitudes about causation that might be taken to constitute a 'folk theory' (Norton 2003, §2.5): that causes temporally precede their effects, that causation is acyclic, that causes make their effects happen.

In practice, only some of the causally relevant factors are identified as causes, others being relegated to *background* (or *boundary*) conditions. The distinction here is grounded in the possibility of intervention, the manipulation of a causal variable in such a way as to isolate its impact on some effect variable. A genuine cause is such that manipulating it can trigger a corresponding difference in value in the effect variable, in a way that manipulating a non-cause cannot (Eagle 2007a, 167). The technique of randomized experiments (Fisher 1935), deriving ultimately from Mill's 'method of difference' (Mill [1874] 1974, bk. III, ch. VIII, §§2–3), is a widespread application of this idea, where the randomization of subjects to treatment and control groups aims to ensure match of causally relevant properties, and hence to enable any treatment effect to be observed in the aggregate impact on the effect variable. Without the possibility of intervention, this methodology is unrealizable. It may be unrealizable in practice anyway, but causal inference techniques rely on the mathematical possibility of intervention so as to 'sever' a causal variable from potential confounders (Pearl 2000, 157–8, 348–51). There is no reason why we can't define a variable whose values are grounded in conditions across a vast swathe of space and time, but obviously such a variable could not have its values arbitrarily fixed by local experimental interventions. Moreover, such variables block the application of standard inference techniques, because their presence in the model potentially undermines the ability to isolate causes from their effects by surgery on a structural equation model. For this reason, the grounds for the values of causal variables are typically localized in space and time.

Even in those sciences where the randomized controlled trial (RCT) is depreciated in favour of *models*, such as engineering or geology, local variables play a key role. The construction of a causal model in those fields is not theory-neutral, as RCTs promise to be, but rests on a particular conception of causal mechanisms as *connections* between variables. The mechanism is a 'black box'; to understand the mechanism is to open the box and examine the detailed linkages between parts of the process within. In this light, a mechanism can be contained, being itself localized and having localizable parts. As Woodward puts it,

> Mechanisms consist of parts, the behavior of which conforms to generalizations that are invariant under interventions, and which are modular in the sense that it is possible in principle to change the behavior of one part independently of the others.   (Woodward 2002, S366; see also Cartwright 1999)

From all this, a picture emerges: the causes involved in the human sciences and many non-fundamental branches of natural science are manipulable, spatiotemporally localized event types; and an ideal causal model in such a science should, while reflecting observed statistical associations, also give us grounds for deciding which interventions are effective in making desired outcomes happen.

2. Effective Strategies and Physical Laws

As many have observed (Norton 2003; Field 2003; Latham 1987; Russell 1913; Cartwright 1979; Frisch 2014; Price 2007), the construction of causal models in the human and special sciences is strikingly different from what goes on in the development of our most fundamental theories in physics. This is evident in both method and product. In method: because of the central role of theory in physical prediction and explanation, the principal direct aim is to discover experimental grounds for the acceptance of theories. Protocols like RCTs, which aim to establish causal conclusions without the intermediating role of models or theories, are thus much less relevant to the epistemic goals of physics. Mechanical models are more theoretically freighted, but even there significant isolating assumptions are required—most notably, robustness to variation in boundary conditions and localized encapsulation of the mechanism. These assumptions are absent in fundamental physics, which tends to have global ambitions.

These observations about method shouldn't be pressed too hard. Causal models, like those constructed via the graph-theoretic approach to causal structure (Pearl 2000; Spirtes, Glymour, and Scheines 2000), are clearly theories about unobserved aspects behind the statistical data. Even more obviously, mechanical models in geology, engineering, etc., make explicit and extensive use of global mathematized physical theory (though under idealizing 'small world' assumptions). Maybe, despite not making any notable use of causal inference algorithms and going far beyond mechanistic reasoning, fundamental physics is also producing theories that are causal, through and through. It would be surprising if this is so, given how important a role substantive causal assumptions are in causal inference—as Cartwright (1989, ch. 2) has it, 'No causes in, no causes out'—and how little evidence there is in physics of any such assumptions being made. Still, perhaps physics uses a different means to arrive at the same causal end.

Accordingly, the main ground for thinking that causal structure is of less importance in physics lies in the content of the theories produced by physics. The theories of classical mechanics, relativity, and quantum mechanics don't involve *localized causal variables* in any obvious way. The don't involve any inference to dependency from coarse-grained associations between the occupants of assorted chunks of space and time. Those theories, quite different in detail, can be nevertheless all be understood very abstractly as involving a complex, perhaps many-dimensional, space (a spacetime manifold, or maybe a higher-order 'configuration space' in the case of some views of quantum mechanics), over which space are given geometric fields characterizing the metric structure of the space, and material fields[2] characterizing its occupants and the values of various physical quantities (Earman 1986, 24; Maudlin 2021; Wallace and Timpson 2010; Ney 2015; Albert 2013). Different models of these theories involve different specifications of the values of these fields across the fundamental space; that which all models of a given theory agree on fixes the laws of a given theory. A theory is thus a family of constraints on the compossible global distributions of field values across spacetime (or configuration space), 'all at once' as it were. These theories don't in any obvious way involve the production of the field values over time. Causation involves making things happen, and nothing in these models is a good candidate 'making' relation.[3]

One obvious reply is that we ought not to look at the theory presented as a 'timeless block'. To do so misses out any *dynamics* in the theory, and doubtless the dynamics are candidate causal relations. In standard formulations, the geometrical structure of the underlying space of a theory will permit the specification of dynamical principles that govern how the material field qualities realized at some region are linked to those at nearby regions. For example, in a geometry which

[2] I am intentionally abstract here, attempting to capture the stress-energy tensor of general relativity and the quantum wave function under a single description.
[3] Sometimes people talk of relations in such theories using causal language: for example, sometimes the light cone structure in special relativity is called its 'causal structure' (Curiel 2021). But on closer inspection this turns out to be 'causal' only in the sense that if spacetime points are spacelike separated (outside the light cone), that is sufficient for them not to be causally connectible of causal relations between them. No causal connections need be instantiated.

permits objective simultaneity relations, there will be instantaneous properties instantiated at times which correspond to rates of change in physical properties over time. The dynamics specifies how these instantaneous rates are linked to the instantiation of the property of which they are rates. For example, velocity is a rate of change of position, and a dynamics will specify how material velocities at some time are linked to material positions at nearby times. The dynamics, presented as a family of differential equations, give rise to the same timeless block model in terms of the global pattern of property instantiation. But the dynamics generates such a block from the bottom up, constraining how the states of different regions can be patched together to form a nomologically coherent whole.

With our interest in causation, the relevance of the dynamics derives from a proposal to understand dynamical laws as causal laws: the local state at some region causing the states that surround it. But despite an etymological affinity, dynamical laws don't fit the profile of properly causal models established in §1. For one, the constraints on stitching states together don't generally have any temporal orientation: a local state constrains its preceding neighbourhood no less than its succeeding neighbourhood. These are 'laws of association' (Cartwright 1979, 419), where the contents of a given region sufficient to fix the contents of other regions will fix its whole surroundings, not just its future surroundings. To get causes here, we need to stipulate that only precedent regions can count as causes.

More importantly, generically these dynamics do not give rise to dependence relations between localized states. Fix on some local state as a putative effect. What generates that local state is enough of its neighbourhood to fix its character, given the dynamics. But this will generally not be itself a localized region. Take the case of relativity, which at least imposes some requirement of locality (through having an upper bound on physical velocity). Any spacelike surface intersecting the past light cone at a point p will fix the character of p; any less will not be enough information for the dynamics to work with to establish what p is like. Such a surface is not typically very localized, and will generally encompass a very large region. Moreover, any such surface will manage to fix the character of the effect. So, dynamically speaking, any arbitrary slice through the past light cone is a cause; all such slices are equally causes; and any proper subregion of such a slice is not enough to be a cause.

If we want an effective strategy for manipulating the contents of a given spacetime point p, and we want to treat the dynamical laws as causal laws describing the operations of our strategy, then the targets of our interventions must be vast regions intersecting every trajectory of potential causal relevance. These include not only what we'd intuitively regard as potential causes, points along those trajectories which might be the paths of genuine causal (mechanical) processes leading to p—we must also specify the *absence* of certain events at any points on trajectories which could interfere with those processes. Such a specification will be completely *undiscriminating*, because every point in the past light cone is potentially occupied by an interfering event. This yields a notion of cause that is not the kind of thing that could be effectively manipulated; these 'causes' are too many, various, and far-flung. The dynamical laws tells us which events we need to specify in order to jointly fix the occurrence of the effect, but these events are not causes, because no humanly effective strategy can exploit or control them (Field 2003, 439). Likewise: take a discriminating recipe that classifies some (but not all) localized spacetime regions as causes. We want causes that 'make the difference'; but intervene on such a region, holding fixed all the other causes, and the dynamical laws will not give any determinate fix on any localized potential effect, because we need also to fix the status of any potential interferers. Dynamical laws can help in identifying effective local interventions only if those local interventions are physically *isolated*, so that there is no potential influence from outside the region of the local cause (Elga 2007). But of course 'being isolated' is not an intrinsic property of a region; to establish that the dynamics can count a localized region as a cause, it needs to be established somehow that the contents of its surroundings are non-interfering. The dynamics alone won't establish this, because they are compatible with models in which local regions are not isolated.

3. Finding Causes in a Physical World

The upshot, I think, is this. Causal relations govern local interventions that effectively (but without guarantee) produce localized outcomes. The models of fundamental physics, even when represented dynamically, don't involve any obvious role for such relations, nor any obvious candidates to be, or surrogate for, them. This leaves us with three main options: elimination, reduction, and supplementation.

Elimination was Russell's preferred option:[4]

the word 'cause' is so inextricably bound up with misleading associations as to make its complete extrusion from the philosophical vocabulary desirable.

(Russell 1913, 1)

[Causal laws] though useful in daily life and in the infancy of a science, tend to be displaced by quite different laws as soon as a science is successful. The law of gravitation will illustrate what occurs in any advanced science. In the motions of

---

[4] See also van't Hoff (2022, §2), who argues that causation must be a discriminating relation, and that 'the insurmountable problem of selection' entails that causation is a 'relation which cannot be instantiated'.

mutually gravitating bodies, there is nothing that can be called a cause, and nothing that can be called an effect; there is merely a formula.

(Russell 1913, 13–14)

Perhaps if our interest lay purely in describing physical reality, Russell's proposal would have some merit. But we ought not 'displace' our use of causal laws to identify effective strategies, in favour of the austere equations of fundamental physics. Even if it were practically possible to do so, many of our effective strategies aren't deemed such by the fundamental laws. The use of mosquito nets is an effective strategy to decrease the prevalence of malaria. This high-level truth won't fall out of the physics unaided; indeed, the pattern of correlations supporting the effectiveness of the strategy won't fall out of physics unaided. (Thus eliminativists who retreat to pure physics can't opt for 'diet causation', i.e., counterfactually robust correlations.) Certainly if we model an individual mosquito net and an individual mosquito as an isolated system, we may be able to predict that the net will block any trajectory of the mosquito that begins on one side of the net and ends on the other. But this relies crucially on tacitly suppressing any possible interferers. Once we embed the mosquito net in the wider world, there will be many physically possible scenarios in which the net is rendered permeable to mosquitoes by some interfering process from outside the narrowly isolated system we were considering. Of course such scenarios are unlikely and not to be taken seriously in deliberation. But that judgement of unlikelihood, and the decision to set aside such scenarios when deliberating, isn't motivated by the physics alone.

Quite generally, suppose we were to try and capture effective interventions in physics by looking at all models in which some local event is realized in some localized region C—the kind of region we might conceivably intervene on in action. Turn the crank on the global dynamics in all of these models, and see what comes out in some target local region of interest E. We would get, in the general case, nothing especially interesting, because the nomological influence of the regions outside C would swamp the influence of C itself. (Here again: unless C is a significant chunk of the past light cone of points in E, the contents of C can be trumped by the contents of the region of the light cone outside C.) To get something out that shows the character of C to fix the character of E, in the absence of knowing the system to be isolated, we're going to have to have some sort of probability distribution over the possible contents of the rest of the initial value surface outside of C (Elga 2007, 118). Such a distribution will allow us to say that, with high probability, intervening on C will fix what happens at E—assuming we assign higher probability to those physical possibilities where no interferers emerge from outside C than to those possibilities containing one or more inter-ferers. Even setting aside the difficulties in constructing a suitable probability function for use in causally-informed decision making—issues I return to below in

§7—nothing in the physics itself will give us such a probability distribution (Norton 2003, 9–10). It might come from our background knowledge, where we just weight the actual world and those worlds similar to it more heavily than other physical possibilities that are more dissimilar to actuality. Or it might come from a general epistemic preference for simpler hypotheses.[5] Neither way does this weighting come from the physics itself, which assigns no special nomological significance to actuality or to any credence distribution over physical possibilities.

This argument that we cannot satisfactorily provide surrogates for effective strategies in the models of pure physics is *a fortiori* an argument against *reduction* of causal laws to physical laws—for reducible causal laws would be perfect surrogates for themselves, assuming again that causal relations are central to the distinction between effective and ineffective strategies. So if we need causal relations to appeal to in deliberation and action, we're going to need to accept that they go beyond the patterns in pure physics. Accordingly, we're going to need to *supplement* physics with additional structure in order to see how it can support causal models.

Obviously we could supplement the physics with enough background information to uniquely specify the actual course of events, perhaps a global initial condition if the physics is deterministic. That would avoid any difficulty with interferers, for physics supplemented with this much information entails the occurrence of everything that actually happened. With this sort of information, however, there is no role for causation; even the idea of an effective strategy is redundant, given deliberation is otiose under such conditions. Prediction on the basis of some event C ought to be guided by the outcomes in some class of physical models in which C obtains (and the differences across a matched population of models otherwise alike in which C does not obtain). The whole class undermines prediction, because almost anything is possible (interferers, the effect coming to pass via some other potential pathway) if we allow arbitrary variation in the non-C-grounding parts of the model. The narrowest class, the model corresponding to actuality, undermines causation because there is no sense to be made of marking C as a cause as opposed to any other condition accompanying a given outcome. So we need information that narrows the class of models in a way that shows the presence and absence of C to reliably indicate a certain effect. What sort of information will do this?

As above, we can add a probability distribution over the possible contents of the environment of a purported cause. This is *modal* information about what is likely possible for the surroundings of our cause. Another source of modal information

[5] The possibility in which there are no interferers outside C is plausibly simpler than the worlds in which there are interferers, so a general Occamist preference for simpler hypothesis might make us weight those possibilities more highly, either in our rational credences, or in some a priori logical probability that weights simplicity strongly (Solomonoff 1964).

that plays a similar role is counterfactual information, and the significance many accounts of causation give to counterfactuals reflects this (Lewis 1973; Hitchcock 2001; James Woodward 2003; Paul and Hall 2013). Counterfactuals concerning what would have been the case had A obtained impose some sort of structure over families of physical models. They divide all A-compatible models into those compatible with what would/might have been the case if it had been that A, and those that, while they are compatible with A, are not consistent with what might have been the case if A.[6] Indeed, even bare modal facts about what is possible for a given system will serve to supplement the physics, as long as they are facts about a species of possibility more restricted than nomological or physical possibility: for example, facts about what a given thing *can* or *has the ability* to do.

That causation requires supplemental facts that are not evident in models of fundamental physics also connects with observations made previously about the role of assumptions of isolation in order that the dynamics might support relations of local determination. When considering a given local region C, such supplemental information will enable us to 'prune' the space of physically possible models that include C. We can eliminate those models in which C's surroundings are too improbable, or its surroundings depart too gratuitously from actuality, or are not surroundings that are realistically possible. That is: this additional information will serve to specify a quite narrow range of permissible assumptions about what the background surroundings of a given potential cause are like, and hold those fixed in the class of models while we 'toggle' C and note the effects of doing so. In the good case, such information will reassure us that 'differences in distant matters of fact are *unlikely* to make a difference' (Elga 2007, 110, my emphasis). And when distant matters of fact are unlikely to make a difference, we can for all practical purposes treat a given system as isolated from its surroundings. Sometimes, of course, the background information tells us that the system isn't isolated, but its environment 'exerts a uniform influence throughout' (Demarest and Hicks 2021, 616). If so, the system is 'quasi-isolated', and the probabilistic and counterfactual information tells us not that the causal region suffices to fix the character of an given effect by itself, but that it manages to make the difference between the effect having one character versus another, relative to that fixed background. When a system can be treated as isolated or quasi-isolated, the dynamical laws and our causal models can happily fit together.

Often, of course, we don't have the slightest idea how to represent the variables of interest in our causal models in the vocabulary of fundamental physics. We might be able to come up with a causal model for the motion of a rock (Elga's

---

[6] This allows us to set aside certain A-compatible possibilities given the laws, for the purposes of deliberation. Of course, setting a possibility aside, and assigning it a significance proportional to its probability, are not the same thing—but for many practical purposes they will yield the same deliberative verdicts.

example) that derives from the dynamics and an assumption of isolation; we aren't going to be able to do the same for epidemiology or agronomy. But the same assumptions of (quasi-)isolation continue to be important, perhaps even more so. In those cases we typically only have statistical data about macroscopic variables of interest, and no microphysical basis. To discriminate causal from non-causal associations we need some background information assuring us that any correlations that are not screened off by some other variable are indicative of causation. We also need some assurance that the causal surroundings permit genuine causes to appear as associations in the statistical data (so they are not masked or interfered with—Hesslow 1976). Both sorts of background information take the form of assumptions that other possible explanations of the statistical data are excluded. And on what grounds do we exclude these other explanations? Typically, if they are too improbable (i.e., a given correlation is statistically significant for some stringent level of significance), or if invoking them would amount to a gratuitous departure from what we think would happen in similar cases (if we need to suppose some sort of 'cosmic conspiracy' in the surroundings to mask the role of a given variable).

Cartwright goes further, arguing that nothing less than information about causal factors will help us discriminate causal and non-causal associations: we need causal information in, to get causes out (Cartwright 1979, 423). Obviously an appeal to a causally homogeneous background will provide modal information that performs the role played in the discussion above by probabilistic and counterfactual information, assuring us that our system is quasi-isolated from other potential influences. Cartwright's invocation of causal information is, in effect, an assumption that the system is quasi-isolated. Then a cause is something that plays a counterfactual or probabilistic difference-making role against a fixed background of other causes.[7]

Let me sum up. Whether the supplementing information is causal or not, the consensus seems to be that causation cannot be located using the resources of physics alone. Information about causal dependencies provides information constraining *what might happen* more tightly than information about the *local* physical state does, even given the background physical laws (though not as tightly as the *global* prior physical state might, especially under determinism). That

[7] Appeals to modal and counterfactual information may also involve causation, though perhaps in a less overt way. For example, not just any counterfactual information helps us delineate the background against which a purported cause makes a difference to the data. Not just any truth that would have still obtained had *P* failed to obtain is part of the background against which *P* is to be evaluated as a potential cause, because if we allow backtracking counterfactuals ('had *P* not occurred, then it would have to have been that *P*' [so as to ensure it is still the case that *E*]) then we end up treating background events as causally relevant. Arguably, drawing this distinction presupposes some causal information, as non-backtracking counterfactuals are 'already assumed to be, in a broad sense, causal' (Hitchcock 2007, 58). Even the notion of isolation itself might be a causal notion: 'A system is isolated (in our sense) just in case all of the features that have causal influence on the suitably specified "output" state of the system go through the "input" states' (Demarest and Hicks 2021, §3.5).

tighter constraint is what makes such information useful for deliberation. (I give more details on *how* causal information plays a role in deliberation in §7.) To try and deliberate using just the resources of physics either requires an appeal to the voluminous body of potentially physically relevant information, which tightly constrains judgements about the outcomes of interventions, at the cost of being both cognitively unwieldy and recommending infeasibly global interventions; or appeals just to local information that provides little constraint on what a local intervention will entail. Moreover, identifying the *kind* of supplemental information that is needed seems to involve ideological resources that go beyond those of the physics: background information sufficient to fix causes isn't physically distinctive, and plays no independently motivated role in the physics (unlike, say, the global background state or the past light cone).

This applies likewise to the information about probabilities and counterfactuals that play a role in grounding the distinction between effective and ineffective strategies: those too cannot be located in the resources of pure physics. Probabilistic or counterfactual dependence also supplements physical law to enhance prediction in the presence of only local physical information. Hence those who agree with Russell that causation is to be eliminated, but who think it can be adequately replaced in deliberation by talk of counterfactual dependence, are equally committed on my view to non-reductionism about effective strategies.[8]

4. The Nature of Supplementation

The upshot of the foregoing is that the ground of a distinction between effective and ineffective strategies, one aligning neatly with our commonplace deliberations, cannot be found in basic physics. Assuming we wish to retain something like our present deliberative strategies, we need to supplement that physics in order to recover the distinction. This idea has precedent:

we all believe in lots of distinctions physics 'can't see'. Arguably,...all [fundamental physics] needs to describe the events with which it concerns itself are things like tiny particles, gigantic fields, and spacetime. Is there no difference,

[8] Dorr (2016, §7) is no causal eliminativist, but he does think 'causal' decision theory is better understood as relying on non-causal counterfactuals; likewise Hitchcock (2013) argues that actual causation plays no role in decision theory, though causal dependence, which is closely linked to counterfactuals, does. Finally, van't Hoff (2022) offers a counterfactual decision theory, tailor-made for the causal eliminativist who wants to respond to the challenges of how to understand deliberation in the absence of causation. Her decision theory, however, relies explicitly on all three of a partition of the physical states into macrostates, a non-fundamental probability function over possible states, and a similarity metric on possible worlds (van't Hoff 2022, §6): her account of credence in a counterfactual $A > B$ sets it equal to the statistical mechanical probability of B in the current macrostate, in the most similar world in which A holds (which might come apart from the actual statistical mechanical probability of B given A and the current macrostate).

then, between groups of particles that make up larger wholes, and groups that do not? Should we conclude that, since physics does not mention things like dogs, there is no reason to believe in such things—as opposed to mere swarms of particles arranged in various canine shapes?   (Zimmerman 2008, 219)

But there are various approaches to supplementation. These approaches differ on the question of the completeness of physics without causal laws. The *moderate* view takes physics to be complete insofar as it talks about physical entities, but maintains that additional facts about causes and causal laws can be added, consistent with the underlying physics. The *radical* view regards physical and causal laws as making incompatible predictions about physical bodies, and hence physics not only needs causal supplementation, but revision.

Both of these approaches have difficulties. The radical view needs compelling evidence that the predictions of physics are inaccurate, and that where causal and physical predictions disagree, it is the causal predictions that are to be favoured. I know of no such evidence. There are plenty of arguments, of course, that we need to idealize and approximate in order to make use of physical models in local prediction and explanation (Cartwright 1983). But these do not amount to arguments that physical laws have empirical counter evidence when the global state is correctly specified. I'll set aside, then, radical approaches.

Moderate supplementers think that physics is correct so far as it goes. A difficulty for this view comes from the further premise that, when it comes to physical facts, 'as far as it goes' is *everywhere*. That is: in principle, when it comes to prediction and explanation of claims about the motion of material bodies, the existence of a fundamental physical model renders other approaches dispensable, even if they are compatible with the underlying physics. Physics is 'authoritative when it comes to predicting and describing the production of physical events' (Eagle 2007a, 174). In part this authority is captured by the claim that everything supervenes on the fundamental; fix how things are physically, and you fix how everything is. (That already blocks certain kinds of 'emergentisms' that would press for a robust autonomy for the special sciences.) But it is not just that everything supervenes on the physical: physics is *complete*: every physical event can be wholly described—and when so described, explained and predicted—in purely physical terms. There is, in principle, an exhaustive non-redundant physical account of every event—and as we've argued, that won't be a causal account. It follows, again in principle, that there is no theoretical requirement for causal explanation of any event.

This is obviously only a very rough and sketchy glance at an argument-type that is familiar from discussions of the autonomy of the 'special sciences' that has spawned a very large literature (e.g., Fodor 1974, 1997; Kim 2005; Loewer 2009; Menzies and List 2010; Robertson forthcoming). My argument has a premise about the redundancy of causal information given fundamental physics, so it doesn't immediately generalize to efforts to defend the autonomy of other special science properties and relations.[9] My argument takes its cue from the manifest difficulty of finding a theoretical role for causal regularities in light of the apparent fact that there are no two possible worlds that are wholly alike with respect to fundamental physics and yet unalike in what causes what. A robust defence of causation would involve finding some way to prevent causal relations being trumped by the underlying physics and nevertheless to acknowledge the proper deference of causal connections between particular events to the underlying nomic relations between global states of the world.

5. The Practical Role of Causal Models

The obvious option is to look, not to the metaphysics of causation, but to the epistemic and practical role of causal models. The key issue that then arises is what to say about the view we ought to take of causal models, if they go beyond physics in this non-metaphysical way.

One proposal would emphasize the role of causation in *understanding*. A non-causal physical model of some phenomenon might fail to be explanatory, in the sense of conveying understanding to the explainee, because whether a scientific explanation conveys understanding depends only partly on its theoretical content. It depends also in part on the conceptual resources available to the prospective explainee. An explanation that is theoretically redundant might nevertheless have utility if the explainee is not prepared to grasp a more fundamental explanation which trumps it. Doubtless causal concepts are more familiar to us than the equations of fundamental physics, and this proposal is surely correct as far as it goes. But for causation to properly ground a distinction between effective and ineffective strategies it seems like we will need a justification for invoking causal relations that is less tied to idiosyncratic conceptual inabilities of individual knowers.

Loewer suggests that the practical and epistemic rationale for special science models is that they

characterize aspects of the structure generated by the fundamental physical laws that are especially salient to us and amenable to scientific investigation in

---

[9] Barring the premise about the all-encompassing nature of fundamental physics, nothing in my argument bears on questions about the autonomy of, for example, mental properties. There may be excellent reasons to suppose that physics isn't complete when it comes to the description of those physical events that turn out to realize mental states, or thermodynamic states. Admittedly, should the grounds for autonomy of the special sciences involve the autonomy of causal explanations involving special science properties, those arguments may be collaterally damaged by my claim for the in-principle dispensability of causation.

languages other than the language of physics.... [T]he autonomy/irreducibility of the special sciences...must ultimately be due to facts and laws of microphysics and to our epistemological situation in the world.   (Loewer 2009, 222)

Loewer suggests that the challenge of the (metaphysical) completeness of physics should be addressed by examining our standpoint. Moderate supplementers can be understood as offering a variety of attempts to clarify how our epistemic situation supports the use of causal models. They have tried to argue that causal models, even if they are in principle unnecessary, are needed in some more practical or derivative way. Two main families of moderates have presented themselves.

Approximators. Some argue that we are justified in making use of causal models because they approximate the real physical explanation under certain conditions. Norton suggests that while causation cannot be recovered from our best science, 'in appropriately restricted circumstances our science entails that nature will conform to one or other form of our causal expectations' (2003, 13). Elga argues that causal models are 'useful to us because so many systems can be treated as isolated from so much of their environments' (Elga 2007, 111). Approximators thus appear to argue that causal models earn their keep through promoting the same goals as physical models—prediction and explanation—and are useful because they can be applied more easily and (under the right circumstances) without appreciable loss of accuracy. But a causal relation that arises only from approximate models is only as real as other entities which are restricted to approximate models: frictionless planes, Newtonian orbits, or Hardy-Weinberg populations.

Perspectivalists. Others argue that causal models serve goals that we have as agents that physical models do not, and that their conditions of adequacy are linked essentially to this agential *perspective*. So, for example, Menzies argues that two physical situations could differ in which causal claims they support while being physical duplicates, due to differences in the context relative to which the causal claims are expressed (2007, 192–3). He ties these differences in context to differences in the course of events that is 'normal or expected or is taken for granted' (Menzies 2007, 220) by the person attributing causation. This context-sensitivity, Menzies thinks, is to be understood as showing that physics needs supplementation by information about such normal courses of events, information that is supplied by our expectations of what will happen if we don't intervene. Menzies tentatively endorses the idea that causal relations are nevertheless real, though in part mind-dependent, given the involvement of expectations. Price (2007) is more full-blooded in endorsing *perspectival realism*, arguing that the deliberative perspective constitutive of our agency, and typical features of what we can know, requires us to make certain assumptions (such as the fixity of the past) that are not required by the physics alone. It is with the support of such assumptions that the outcomes of our interventions are established (Price 2007, §§10.7–10.8). Again, a mind-dependent feature, peculiar to deliberators constituted in roughly the way that we are, is understood to be part of what grounds certain causal claims. Finally, I also endorsed a kind of perspectival realism (Eagle 2007a, 185–7), tied to another aspect of our agential perspective: that we seek understanding and the concomitant ability to generalize our explanations to new scenarios (typically by representing our explanatory conclusions as counterfactual judgements). To do this requires abstracting away from details that might be highly relevant, physically speaking, and adopting a modally rich perspective that involves, once again, certain perspective-dependent assumptions about which details are important, and which are dispensable.[10]

I once found the perspectival approach deeply appealing. It appears to hold out the promise of both respecting the completeness of physics as a theory of things given 'the view from nowhere' (Nagel 1986), while simultaneously supporting the robust reality of causal relations, from the embedded perspective of agents with a particular epistemic position and a circumscribed sphere of influence. It would neatly thread the dilemma we now face. If causal notions aren't fundamentally real, but only real 'from our perspective', we needn't embark on any revision of fundamental physics. But if being real from a perspective is a way of being real, we might have a notion of causation that is real enough to vindicate the causal models at the heart of the human sciences. This last contrasts with the approximator's position that our causal models merely approximate reality, without being in fact accurate; for those who want a more robust independence of the human sciences from any physical basis, the perspectivalist view holds more appeal. But we must be wary of confusing a wish that perspectivalism be true, with an argument that it is.

Causal perspectivalists have some unfinished work to explain exactly what 'real from a perspective' is supposed to come to (Schaffer 2010, 848). One interpretation takes the view to be unashamedly metaphysical, a view perhaps close to Fine's 'relativism', the view that 'Reality is relative to a standpoint; and for different standpoints there will be different realities' (Fine 2005, 262). Such a version of perspectivalism will deny that fundamental physics is authoritative: physics might give us non-relative aspects of reality, but it will omit precisely those facts, like causal facts, that emerge only given an agential perspective. Yet the central conceit of 'relative facts' is hard to accept, especially in this case where (unlike other cases where relativism has been defended, such as tense or the first-person) there is no obvious non-analogical sense of a standpoint being invoked. It would be better to be able to do without radical metaphysics; or perhaps, if one is

---

[10] The approximator/perspectivalist distinction may not be especially hard-and-fast—Norton (2003, 14) says that causes have a 'derivative reality', not so different from the kinds of things perspectivalists say.

going to opt for a radical view, a radicalism that is tailored to causation (like that mentioned at the start of this section) seems better motivated.

But this reading of Price's perspectivalism is not perhaps the most plausible; it might be more charitable to read him as committed, like Menzies is, to a kind of contextualism about *cause*. (Indeed, prominent interpretations of Price, like Ismael (2015), read him as committed to an explicit thesis that *cause*, like *local* or *near*, has some covert argument place filled in automatically by a contextually given perspective.) The contextualist argues that the word *cause* is context-sensitive, so that '*A* causes *B*' could express, relative to context *c*, something like '*A* raises the *c*-expected probability of *B*'. The role of context here is to supply just the supplemental ingredient we mentioned at the end of §3: namely, probabilistic or modal information. Contextualism has an advantage over theories which require explicit calculation of the relevant probabilities, if only because context can supply the probabilities to truth-conditions directly without awareness on the part of the speaker. But this form of contextualist perspectivalism, while non-mysterious, doesn't address the central worry that the propositions expressed are redundant given the physics: we still need some argument as to why the propositions expressed in context by causal claims can play an interesting role in our conceptual economy given their fundamental dispensability.

In my earlier paper, I suggested that a *fictionalist* approach might clarify perspectivalism:

> utterances within a given perspective, just as utterances in a fiction, are to be interpreted semantically as if the content of the fiction were true, but not as representing that the content obtains actually.   (Eagle 2007a, 188)

On re-reading my earlier discussion, it is far from clear how exactly I thought fictionalism related to perspectivalism, and on the face of it, there is a significant tension between the non-committal ontology of the typical fictionalist and the metaphysical radicalism of the perspectival realist. It is perhaps easier to see how contextualist perspectivalism is compatible with fictionalism. But as contextualism is a theory of truth-conditions—not of whether those conditions obtain—it is difficult to see how contextualist perspectivalism can in fact fulfil the perspectivalist promise to reconcile the completeness of physics with a robust commitment to causation. Perhaps there is a contrast between the context-insensitive language of fundamental physics, and the context-sensitive language of causation—but something more needs to be said about how that contrast could illuminate the metaphysics of causation.

However, while I'm less persuaded than my earlier self of the promise of perspectivalism, I continue to favour fictionalism as an approach to understanding the relation of causation to fundamental physics. In the remainder of this chapter, I want to try and develop the very preliminary remarks I made in

2007 a bit more, and really explain and attempt to justify a fictionalist approach to causation.

Before leaving other moderate supplementing approaches, I want to suggest that fictionalism may be helpful in understanding the approximator's position also. For some approximators may wish to endorse the idea that causal models are models, in the same way that an ideal gas or a frictionless plane is a model: entities that are known to be unreal, but that are used to represent a target system with which they share certain explanatorily potent features. There is a considerable appeal to the idea that scientific models are like fictions, 'creatures of the imagination' as Godfrey-Smith (2009, 101) puts it (see also Contessa 2010; Frigg 2010). I will explore a fictionalist account of causation with some awareness of this potential application, but primarily with the goal of trying to offer an account of our construction and use of causal models that might vindicate our practice in the absence of a reduction to physics (or a Cartwright-style causal inflation of physics).

6. Fictionalism

There are many varieties of fictionalism (Sainsbury 2010; Eklund 2019; Kroon 2011; Caddick Bourne 2013). The fictionalism that interests me is an interpretative approach to an apparently entity-involving discourse D which

> claims that those participating in D should not have (and perhaps, in the case of certain types of discourse, typically *do* not have) truth as their aim when they accept a sentence from D. The norm for acceptance is not truth—one can accept a sentence of D for the best of reasons, justifiably acting on it, using it in one's theorising, drawing inferences and acting on those, and so on, but not actually *believe* it, since there are benefits other than truth for whose sake one should accept such a sentence.   (Kroon 2011, 787)

This kind of view gets the name 'fictionalist' because this is an interpretative stance that can often be taken towards fictional discourse. Take the case of fully participatory engagement with fictional texts (Eagle 2007b, 128): for example, in a seminar on *Bleak House* in which people describe people's actions and speculate on their motivations without hedges like 'according to Dickens' or 'in *Bleak House*'. It is implausible to suppose that these hedges are nevertheless there, covert and unpronounced. Fictionalism about fictional discourse ought not to be pursued using what Lewis termed 'disowning prefixes' (Lewis 2005, 315). That sort of account makes incorrect predictions about how participatory discourse about fictions embeds (R. Joyce 2005, 292–5), and how it interacts with our attitudes (Yablo 2001, 76), and about the flexibility of the conversational role of such utterances (Eagle 2007b, 131–3). So it is preferable to take the content of the sentences at face value.

Instead, fictionalism ought to be pursued by some strategy that allows these sentences to be uttered without being assertions, while looking for all the world as if they are. If there is an explicit 'disowning preface' that signals an entire body of utterances isn't to be taken to have the force of assertion—such as Lewis' example, 'I shall say much that I do not believe, starting *now*'—then we needn't look any further for grounds for fictionalism about those utterances. Sometimes the preface only conventionally implicates that the speaker disowns what is to follow; that, I take it, is the function of 'once upon a time' as a traditional opening to fairy tales.

Most discourses lack any obvious disowning prefix.[11] Fictionalists about a given discourse typically need to figure out *why* someone might disown the entailments of what they say, in the absence of any preface assuring us that, for some reason, they do. Hence fictionalists have tended to attempt to come up with reasons why a given discourse might be worth pursuing in absence of any commitment by speakers to its content. These might be understood as reasons we might have for retaining the discourse once we discover it to be erroneous, or might be understood to be the reasons we've all along implicitly had for adopting that form of discourse in the first place.[12] Either way, identifying such a purpose for a given discourse is to try to come up with a standard for good and poor contributions to the discourse that is distinct from the truth-normed standards of non-fictionalistically interpreted discourse. Perhaps in the case of participatory criticism in the classroom, the purpose of talking that way is to foster imaginative engagement with the work without the distancing that explicit mention of the fiction would encourage. The relevant norm is not to assert truths, but to make your utterances those that facilitate the achievement of this purpose, by allowing yourself and your classmates to suspend disbelief and bring their considerable skills in folk interpersonal psychology to bear on the fictional situation.

Fictionalism about a domain of discourse isn't committed to thinking that the norm of that discourse is the norm of fictional discourse. There are many possible purposes for a practice, including linguistic practices. Fictionalists needn't think that fictionalism about a domain of discourse has to involve taking its point to be the same as a discourse engaging with a work of fiction (Caddick Bourne 2013, 148). So understood, fictionalists needn't discuss in detail the metaphysics of fictional entities or the structure of operators like 'in *Bleak House*...', except of course insofar as such discussions are helpful in generating suggestions about how to understand their target domain of discourse. Similarly, while the attitude

---

[11] Lewis (2005, 319), however, thinks that even raising anti-realist metaphysical hypotheses about *F*s prior to engaging in apparently *F*-committal discourse can suffice for disowning.

[12] This is close to the distinction between *revolutionary* and *hermeneutic* fictionalism (Burgess and Rosen 1997; Stanley 2001).

fictionalists recommend taking to a given discourse must fall short of belief, it needn't be restricted to those attitudes, like *pretending*, that have been invoked in the service of understanding fictional discourse.

A further important way in which fictionalists distance themselves from paradigm cases of fictional discourse is in any commitment to an *error theory* about the discourse. There need be no commitment to the falsity of any distinctive claims in the discourse. (In fact, even fictionalists about fiction might wish to reject an error theory for historical fictions, which may contain many true historical details, or certain instances of postmodern 'metatextual' fictions, which may well say true things about themselves.) Some fictionalisms may be motivated by a conviction that some central parts of the discourse cannot be true—moral fictionalists, for example, may be convinced that no properties could function in the way moral properties are expected to. But to say of a discourse that any assertion-like speech acts it involves don't implicate belief, isn't to say that one should believe the contents of those acts to be false.

Given that we have distanced fictionalism from any requirement that distinctively 'literary' phenomena be involved, and we are treating discourses where, semantically, any disavowal of the content is at most present in a tacit preface, fictionalism ends up having a fairly thin characterization. To be a fictionalist about *F* involves offering an orthodox semantics for *F* discourse that is continuous with the rest of the language, coupled with an unorthodox account of the norms governing assertion-like speech in *F* discourse. Fictionalism is thus a fairly broad church. But without any recourse to the conventions surrounding fiction (like filing a book under 'Fiction' in the library), or to the standard fictional attitudes of pretence or make-believe, would-be fictionalists are confronted with several questions about their approach to *F* discourse.

Disavowal. Why should we treat *F* discourse as not requiring truth to be successful, nor as requiring its users to believe in its content?

Retention. Why retain that discourse and continue talking of *F*s, rather than stick with truth-normed discourse?

Aim. What is the purpose of the discourse in the absence of an aim at truth?

Attitude. What attitude should participants in the discourse take or be understood as taking? What is the assertion-like speech act involved?

Answers to these questions will constitute a characterization of a species of fictionalism about *F*s, and an argument for adopting fictionalism. If we can provide satisfactory answers to the first two questions, that provides a *prima facie* case for a fictionalist approach, at least in the presence of the auxiliary assumption that talk of *F*s should be given a semantics continuous with the rest of natural language. The ultimate case for fictionalism also requires that there be plausible answers to the second two questions. A plausible fictionalism will invoke attitudes that are psychologically realistic, speech acts that are at least arguably attested in other domains, and an aim that is a legitimate goal for a human practice. The questions aren't necessarily to be taken sequentially, however; if we cannot give a reasonable non-cognitivist account of the aim and attitudes involved in a discourse, we might well revisit the question of whether we ought to retain or disavow the discourse in the first place.

Proponents of fictionalism about various subject matters have been more or less dutiful in attempting to answer these questions. So moral fictionalists (Kalderon 2005; Joyce 2005) might say: we disavow moral discourse because of its queer metaphysics; we retain moral talk because precommitments to 'exclude from practical deliberation the entertainment of certain options' (Joyce 2005, 307) are an effective way to bolster the will to avoid socially and individually detrimental temptations; the purpose of moral discourse is to establish such precommitments; and we take an attitude to a moral claims M which is approximately *deferring to* M *as a guide in action*. Joyce thinks this last attitude is a kind of collective make-believe, where we choose to engage with the fiction of morality in order to establish certain dispositions in ourselves to react in 'the moral way'. Many have questioned whether engaging with an explicit pretence does manage to inculcate the relevant dispositions.

Closer to our concerns is the proto-fictionalism about theoretical entities espoused under the name 'constructive empiricism' by van Fraassen (1980). His answers to our guiding questions are something like this, on my interpretation:

- We needn't endorse (by believing) the content of theoretical entity claims because empiricism ensures that such claims could not amount to knowledge (which norms assertion).
- We ought nevertheless to retain theoretical entity discourse because being 'totally immersed in a scientific world-picture' (van Fraassen 1980, 81) is vital for the successful pursuit of scientific activity, both experimental and theoretical. Science proceeds best when scientists work with theories having claims about unobservable entities as part of their content, rather than working with bowdlerized theories only mentioning observables.
- The aim of science is the construction of 'empirically adequate' theories (theories that are accurate in what they say about observable phenomena).
- The attitude involved is *acceptance*, amounting to a belief that a theory is empirically adequate,[13] coupled with a practical decision to act more or less as if one believed the theory. The associated speech act seems to be something like assertion within the context of supposing the content of the theory:

---

[13] While the attitude of acceptance involves belief in what's said by a sentence with some kind of disavowing prefix—for example, 'So far as empirical adequacy is concerned, *P*'—the content of what is accepted is just *P*.

> Even if you do not accept a theory, you can engage in discourse in a context in which language use is guided by that theory—but acceptance produces such contexts.   (van Fraassen 1980, 12)

A satisfactory fictionalist theory of causation will answer our four questions. Most of these questions can be addressed by taking up points already made in the first half of this chapter. I particularly wish to revisit those earlier arguments in light of van Fraassen's brand of fictionalism. It would not be wildly inaccurate to characterize my proposal here as a kind of constructive empiricism about causation, but where acceptance of causal models is keyed to our deliberative goals rather than scientific inquiry more generally.

We've rehearsed at length the reasons to think the causal talk cannot be given a fully realist treatment: physics is complete (§4), but nothing in the physics corresponds to the relations of local determination characteristic of causation (§§2, 3). Nothing in these observations tells us that causal models are incompatible with the underlying physics, so we don't have an argument that we have to give up causal representations of the phenomena. But the completeness of physics does suggest that, equally, we cannot be required as a matter of theoretical adequacy to include causal relations in our global models of reality. Representing things causally is optional. I'll argue in the next section that using causal representations is an option we should exercise; and in the final two sections (§§8, 9) that the best way to balance the retention of causal thought and talk with their optionality is to treat them as a 'useful fiction' (cancelling, of course, the implication that disavowing a causal representation involves regarding it to be false). The point of causal representation isn't to duplicate the function of fundamental physics; and fulfilling the function of fundamental physics isn't all we want from a model of reality.

7. Retaining Causal Talk and Models

Nevertheless, we ought to retain causal talk and our deployment of causal models in situations of human interest. Our earlier discussion also showed that physical models don't give us a good handle on the distinction between effective and ineffective strategies (§2). Interrogate the physics for a potential intervention with respect to some local outcome, and you get back a vast region containing every event of potential physical relevance to that outcome. Changing the character of the whole region is guaranteed to affect the outcome, but is practically infeasible. But the physics gives no guidance about which subregions are the most potentially effective loci of action. But that is precisely the information we want, and that causal models give us.

In Eagle (2007a, 177), I emphasized the importance of causal information for explanation. The way I'd now put the idea is this. The vast network of events of potential physical relevance is altogether too complex and detailed to be explana-tory. This is not just because our understanding is too limited to grasp it, but because explanation itself is essentially an abstractive process. To explain an outcome is not to show that its physical antecedents necessitated it. Rather, we explain X when we answer a contrastive question: *why X rather than some salient alternatives* X', X'',... (van Fraassen 1980, 127). The answer we give must cite some event on which the occurrence of X rather than X' depends: some event such that, had it not happened, some alternative to X would have occurred. Explanatory dependence appears to be a species of counterfactual dependence.

But counterfactual dependence may not suffice, as cases of Simpson's paradox show. Suppose we are considering some statistics about college admissions. The real statistics are a bit more complicated (Bickel, Hammel, and O'Connell 1975), and I'm going to tweak the variables of interest to illustrate my point more clearly. Suppose we collect statistics on admission broken down by whether the applicant applied in the main round to start in the autumn, or for mid-year entry, starting in the spring. We see something like these statistics: overall, the chance of a mid-year entry applicant being admitted is 12.5%, while the chance of a main-round applicant being admitted is just 10%. This might suggest that a relevant explana-tory factor of an applicant's chance of admission is time of application: that a main-round candidate might have had a higher chance of admission had they applied for mid-year entry instead. This might also suggest certain hypotheses: for example, that admissions committees favour mid-year applicants. But before jumping to policy recommendations, we look at the raw data in a more fine-grained way, broken down by department of application, as in Table 9.1.

Table 9.1 Imagined admissions data

	Admit/Apply: English	Admit/Apply: Spanish	Overall
Mid-year	0/15	5/25	5/40
Main round	6/135	10/25	16/160
Overall	6/150	15/50	21/200

What we see is that, though overall main-round applicants have a lower admission rate than mid-year applicants, they have a higher admission rate in each department to which they apply. These data suggest many hypotheses about why mid-year applicants apply in this distinctive way, but the hypothesis that they are favoured by admissions committees cannot be sustained. The driver here is that mid-year applicants apply preferentially to more competitive departments: most main-round applicants apply to English, with a 4% admission rate, and most mid-year applicants apply to Spanish, with a 60% admission rate. This additional information then suggests an opposing counterfactual: that had a main-round candidate applied mid-year, they would have had a lower chance of admission.

These counterfactuals have contrary consequents; which is right? Arguably, both—*in the appropriate contexts*. Make the context explicit, and we see that both of these counterfactuals are true:

(1) Had the applicant applied mid-year, they would have had a higher chance of admission—*because had they applied then, they would have more likely applied differentially to less competitive departments.*

(2) Had the applicant applied mid-year, they would have had a lower chance of admission—*because mid-year applicants have lower admission rates, holding fixed the department to which they actually applied.*

Both counterfactuals might be true. But (2) is explanatory in a way (1) is not. This is because it holds fixed a causally relevant factor. As Cartwright notes in her discussion of the real data,

> The difference between the two situations lies in our antecedent causal [assumptions]. We [assume] that applying to a popular department (one with considerably more applicants than positions) is just the kind of thing that causes rejection. . . . If the increased probability for rejection among [main-round applicants] disappears when a causal variable is held fixed, the hypothesis of discrimination in admissions is given up . . .   (Cartwright 1979, 433, slightly altered)

By contrast, though (1) is true and tracks a genuine statistical association, it focuses on a feature—being a mid-year applicant—which we intuitively don't regard as causally significant. We may think that a fixed applicant is likely to have 'intrinsic' preferences between disciplines, but that the question of whether they apply mid-year or not is a more external and haphazard matter. There is no natural kind, 'mid-year applicant'—by contrast, 'English applicant' seems a more natural category.

If this is right, then good explanations will invoke counterfactuals that are appropriately backed by causes. It will not do to explain why a given candidate was admitted by citing their status as a mid-year applicant. A better explanation cites the department to which they applied, and recognizes that differences between admission rounds in the distribution of applicants to disciplines may play some role, but it is not a straightforwardly causal one. Causation is here essential for discriminating better explanations from worse, even in a broadly counterfactual approach to explanation.

What goes for explanation also goes for deliberation. An agent deciding what to do must balance the expected costs and benefits of various actions. Suppose an applicant is deciding when to apply to graduate school; having just missed the deadline for the main round, they slightly prefer to apply mid-year, rather that wait to the following year. But they vastly prefer being admitted to being rejected.

This might summarize their preferences over various outcomes (act/state pairs), assigning numbers—*utilities*—roughly to track something like the degree of relative preferability, as in Table 9.2.

Table 9.2 Admission decision: matrix of value

Acts	States	
	Admitted	Rejected
Apply mid-year	49	1
Apply next main round	45	0

Evidential Decision Theory (EDT) says that, given these utilities for outcomes, you ought to choose the action that has the best subjective expected utility. The subjective expected value of an act's utility ('subjective expected utility' or *SEU*, for short) is the credence-weighted average utility of the act across all states. Because the credence of a state isn't independent of the act, we need to use the conditional credence of the state given the act. The general framework is this. Given a potential act A, some background states $S_1, ..., S_n$, someone's utility function U, and their credence function P (a probability function), the expected utility of A is defined:

$$SEU(A) \stackrel{\text{def}}{=} \sum_{i=1}^{n} U(S_i \wedge A)P(S_i \mid A).$$

These numbers lead to decision in accordance with this proposed norm:

EDT   You ought to perform the act that has the highest subjective expected utility of those open to you.

What the admissions example shows is that this norm goes awry. Applied uncritically, we might reason as follows:

We are trying to establish the relative merits of the acts *apply mid-year M* and *apply next year N*, with respect to the outcomes of being admitted (A) or rejected ($\neg A$). The expected utility of applying mid-year is

$$SEU(Y) = U(A \wedge M)P(A \mid M) + U(\neg A \wedge M)(1 - P(A \mid M))$$
$$= 49 \times 5/40 + 1 \times 35/40 = 7.$$

The expected utility of applying in the next main round is

$$SEU(N) = U(A \wedge N)P(A \mid N) + U(\neg A \wedge N)(1 - P(A \mid N))$$
$$= 45 \times 6/160 = 4.5.$$

So we ought to apply in the mid-year round.

This looks like bad reasoning in light of the fact that in each department, applicants have better prospects in the main round. The overall statistics are correct, as far as they go, but they are misleading because the increased in probability of mid-year admission is an artefact of the way the statistics are analysed. As Cartwright noted, we ought to think not about whether mid-year application is *correlated* with admission, but about whether it *causally promotes* admission.[14]

Cartwright does not offer a decision theory, exactly. She does offer a definition of 'effective strategy' (1979, 431): S is an effective strategy for G if the expected conditional probability of G given S is higher than the expected unconditional probability of G, where that expectation is calculated over a partition of the outcome space by the causally relevant background factors. Broadly following Skyrms (1980), this idea can be incorporated into a decision theory. This will be a version of *Causal Decision Theory*, though of a sort not so popular these days.

Let $C = \{C_1, \ldots, C_n\}$ be a *causal partition*: a division of the outcome space into mutually exclusive, jointly exhaustive cells, such that each C_i specifies some way for the relevant causal background factors (those not at the time of decision under the agent's influence) to be distributed. The partition will be one that an agent regards as a good way of thinking about the causal structure, in light of how they represent the situation. There's no requirement that the partition reflect the 'real' causes, whatever those might be. Rather it reflects factors that this agent regards as the background against which they might act. Lewis (1981, 11) suggests that the members of this partition for a given agent could each be a 'maximally specific proposition about how the things [the agent] cares about do and do not depend causally on [their] present actions'. Skyrms (1980, 133) gives a similar characterization: the cells should be 'maximally specific specifications' of the factors outside the agent's influence (at the time of decision) which are causally relevant to the outcomes. I take it that these descriptions will be satisfied if the cells of the partition are determined by whatever causal model of the decision situation is

---

[14] Some versions of EDT (Price 1986, 199; Ahmed 2014) counsel against using these statistics in setting one's credences, and instead advise using the same statistics used by Causal Decision Theory, as below. Whether these versions of EDT are 'really' non-causal, given the reasoning they deploy constraining the choice of appropriate probability function, and the deflationary version of CDT introduced below, is an interesting question.

available to the agent. Given such a partition, let the C-expected credence of O given A, relative to C, be defined:

$$P_C(O|A) \stackrel{\text{def}}{=} \sum_{(C_i \in C)} P(O|A \wedge C_i)P(C_i).$$

With this notion in hand, the needed change is simple: agents shouldn't use their credences in outcomes given acts in deliberation: they should use their C-expected credences. The way to implement this is to define a new, causally sensitive notion of expected utility, to sit alongside *SEU* (it's just a definition, so can't be right or wrong). The change we need is in our norm; rather than use *SEU* to evaluate possible acts, we should use this new notion. So let's define the *subjective causally expected utility* (*CEU*) by adapting our earlier definition of *SEU*:[15]

$$CEU(A) \stackrel{\text{def}}{=} \sum_{i=1}^{n} U(S_i \wedge A)P_C(S_i|A).$$

In general, of course, $P(O|A) \neq P_C(O|A)$. Then proposed norm is:

CDT   You ought to perform the act that has the highest subjective causally expected utility of those open to you.

Apply this to our admissions example. The background causal partition in our simple model is given by which department an applicant is to apply to: English (*En*) or Spanish (*Es*). Presumably it was once under the agent's control to choose a major; having done so, they are no longer in any position to make a choice about which department to apply to. (This is vital, of course: for it blocks the agent from changing their intended department and thus exploiting the statistical quirk that makes mid-year entry look *prima facie* appealing.)

The relevant probabilities can all be extracted from the frequencies presented in Table 9.1:

$$P(En) = 150/200 = 3/4; P(Es) = 50/200 = 1/4.$$

Hence:

$$P_C(A|M) = P(A|M \wedge En)P(En) + P(A|M \wedge Es)P(Es)$$
$$= (0/15 \times 3/4) + (5/25 \times 1/4) = 5/100; \text{and}$$
$$P_C(A|N) = P(A|N \wedge En)P(En) + P(A|N \wedge Es)P(Es)$$
$$= (6/135 \times 3/4) + (10/25 \times 1/4) = 2/15.$$

---

[15] This definition is very close to one of Lewis' reformulations of his causal decision theory (1981, 15).

Now we can calculate *CEU* for our possible acts:

$$CEU(M) = U(A \wedge M)P_C(A\,|\,M) + U(\neg A \wedge M)(1 - P_C(A\,|\,M))$$
$$= 49 \times 5/100 + 1 \times 95/100 = 3.4;\text{ and}$$
$$CEU(N) = U(A \wedge N)P_C(A\,|\,N) + U(\neg A \wedge N)(1 - P_C(A\,|\,N))$$
$$= 45 \times 2/15 = 6.$$

The final step is to apply our new causal norm CDT, which (in disagreement with EDT) recommends that you ought to apply in the main round, regardless of your chosen subject. The credences involved in rational decision should be causally informed, not merely reflecting the statistics, but reflecting how you as an agent are positioned with respect to those statistics. The bare fact that mid-year entrants have a higher success rate *simpliciter* doesn't reflect the decision facing the agent. They are trying to decide what to do, holding fixed what they are assuming is no longer under their control, if it ever was—such as to which programme they might be applying.

Most recently, popular versions of Causal Decision Theory follow Gibbard and Harper (1978) in calculating expected utility using credences in causally inter-preted counterfactuals (Lewis 1981, 21-8; Joyce 1999; Hitchcock 2013), rather than using causally informed credences. The counterfactual approach is, however, beset by various controversies and challenges, about the logic of counterfactuals and their interaction with conditional probabilities, that the causally informed credence approach neatly sidesteps. The causal partition is, of course, rich in information that entails counterfactual claims (or at least, constrains the proposi-tions they express relative to various contexts), and it may well be possible to extract the Gibbard and Harper version of Causal Decision Theory from the Skyrms/Lewis/Cartwright theory. The version of Causal Decision Theory sketched here has the virtue of making the role of causal information and causal models especially prominent.

This has been a long excursion through the details of causal explanation and causal decision making. The details matter, because they show that appeal to causal information is unavoidable if we want theories of explanation and rational decision that give correct verdicts about what to think or do in various clear cases, and hence what might be a viable candidate to extend as a general purpose approach to explanation and deliberation. The overall argument for retention of causal notions is that our notions of rational choice and explanation are thor-oughly permeated by irreducible causal notions. Causal information, drawn from our pre-theoretical background assumptions or (preferably) from properly tested causal models, is a precondition for explanation and deliberation to proceed in anything like the way we think they do. Conversely, if there were no agents in need of understanding, or no one needing to make choices, there might be no need to invoke causal models for any other purpose. Causal information is an essential precondition to taking a deliberative stance, but nothing in the physical laws necessitates that any agents take that stance.

The role of causal information in choice and explanation lies somewhat in the background. In the version of Causal Decision Theory sketched above, it plays a role in the delineation of a causal partition. In the counterfactual theory of explanation, it plays a role in characterizing which contexts are such that true counterfactuals in them are explanatory—namely, those contexts in which the information held fixed under counterfactual assumptions is causally relevant to what needs explanation.

In neither case is the truth of the causal information at issue in the deliberative or explanatory project.[16] In deliberation, what is centrally at issue for an agent is what to do; their decision turns on the actions available to them, and the impact of those potential actions on the outcomes they value. The framework above involves the agent antecedently making causal assumptions about how their actions lead to outcomes. But what is at issue in a claim about what it is rational for someone to do—what two rational agents might be centrally disagreeing about if they reach different verdicts about the right course of action—is the values they espouse and the credences they have. The broadly presuppositional role of causal information in explanation and deliberation means that the truth or falsity of causal claims is largely beside the point for agents. Having causal presuppositions is enough to get the deliberative project off the ground; those presuppositions provide a framework for having an agential perspective.

Eliminativists about causation, like Russell, will deny the truth of these presuppositions, and argue that the deliberative project ought to proceed in some radically different way, relying only on physical structure, and not on any proposed 'excess' causal structure. I see no real prospect that this reconstructive ambition will be satisfied. Consider an attempt to pursue it within the decision theory sketched above. The background partition cannot be a causal one any longer, but must rather include background states of physical relevance that are not under the agent's control. There are only two ways to proceed here.

- We can include every physically relevant background condition. This will give us an extremely fine-grained partition; indeed, under determinism, the partition will assign each possible physical model its own cell. This will enable us to define a notion of physically expected credence—*if* agents have credences over this extremely rich partition. Even if that unrealistic assumption is met, the partitions are so fine-grained that the background partition, for any proposed action A, already entails A. The physically expected credence of an outcome given an action reduces to just the expected credence of that outcome. When this credence is plugged into the *CEU* formula, we get the triviality that the expected utility of an action is just the expected value of a physical state—action falls out of the picture entirely, and this is no longer a theory of deliberation (rather it is a theory about what physical process you should hope take a course through your body and environment). As Price (2007, 281) puts it: a deliberator must think of 'her own actions as...not themselves determined by anything "further back"'. The fine-grained physical partition is such that each cell entails an action (or its negation). Accordingly the agent who uses this partition isn't really deliberating, on Price's view: they have credences in which bodily movements they'll come to perform, and they have hopes and wishes about those prospects, but they can't be deciding, as that involves the agent making it such that a certain action occurs, as it were 'independently' of its physical precursors. The deliberator whose background partition is fine-grained enough won't be able to ignore the fact that this picture of choice simply doesn't seem to be reconciled with the existence of a complete physical account of each bodily movement.[17]

- If we don't include every physically relevant background condition, we are back to the original admissions case with which we began, where spurious associations between variables which are physically real must be regarded as candidates to guide credence in decision, as in EDT. The existence of intuitively 'better' choices of background variables is irrelevant in the absence of physical grounds to privilege those variables. If we somehow try to make use of all possible selections of potential background information, our decision theory will be worryingly non-committal about which actions are rationally preferable.

Neither of these approaches is satisfactory for us. Either we aren't offering an account that helps us make decisions, or we offer an account which is subject to counterexample. The invocation of causal information makes for a useful and plausible decision theory.

The case for retention of causal models in the face of their redundancy in the face of physics boils down to the fact that for agents like us, causal models provide non-redundant background information that is vital for intuitively defensible choices and explanations. The content of the 'causal fiction' is something like this: that there is a privileged way of partitioning the space of outcomes that

---

[17] Van Fraassen makes the related point that an omniscient being is not in the business of seeking explanations; there is necessarily some ignorance involved in any explanation—as well as some background assumptions taken for granted which determine the kind of explanation sought (van Fraassen 1980, 130).

reflects the background causal structure. Adopting this fiction overtly, or implicitly presupposing it, is a necessary precondition for us to do anything that resembles deliberation and explanation.

This case for retention of causal notions is basically that we couldn't rationally deliberate without them. But we also need some guarantee that rational deliberation is a viable strategy; that our world isn't fundamentally hostile to limited agents like those we take ourselves to be. The case that the world is *approximately* causal would provide such a guarantee, vindicating a deliberative stance as a good trade-off between accuracy and implementation. We discussed part of this case in §3 when discussing the fact that, by and large, many physical systems can be treated as if they were (quasi-)isolated, and hence approximated by a local causal model.

8. The Aim of Causal Models

According to van Fraassen, there is no prospect of empirical knowledge of theoretical entities. Norms governing assertion then forbid us from asserting claims about theoretical entities. What then is the rationale for continuing to make, in an assertion-like way, claims about theoretical entities? Van Fraassen suggests: full immersion in theoretical entity discourse facilitates achieving the aims of the science—namely, the development of empirically adequate theories. Van Fraassen is, on my reading, a causal fictionalist, as causal relations are unobservable, posited to explain observed correlations and patterns. We could follow van Fraassen and argue that the aim of a causal model is just the aim of science more generally: empirical adequacy.

This wouldn't be a wholly satisfactory answer for us. For one thing, the arguments above don't involve anti-realism about theoretical science; causal fictionalism is compatible with robust realism about physics. For another, as just discussed at length, causal models seem to play a distinctive role in our cognitive economy. They are retained because they help us not with 'science' in general, but with particular applications of scientific knowledge. A good account of the aims of causal models, and the human sciences more generally, would explain how invoking causal notions in deliberation and explanation conduces to that aim.

The obvious thought given what we've already argued is that the aim of constructing causal models is to enable effective deliberation and explanation. I've already suggested that without causal assumptions, we may not even get activities that are recognizable to us as deliberative. So in order to facilitate our competent participation in an activity that is fundamental to limited agents like ourselves, we need to make assumptions about the background fixed causal structure against which we consider our options and evaluate candidate explanations.

Our causal models need not be true to play this role. But even very modest externalistic constraints on explanation and deliberation must say that rational choice and successful explanation cannot float wholly free from worldly matters. (Perhaps some literary fictions are measured by wholly internal standards of achievement, with their aims not involving truth even in part, but I don't think the causal fiction could be like that.) Just as the constructive empiricist says that successful science should be true in what it entails about observable matters, the causal fictionalist should, I think, say that successful causal models ought to be *physically adequate*: true in what they entail about physical matters. The causal relations they involve is excess structure from the perspective of physics, but the correlations and patterns of occurrence between variables that a causal model entails should ideally agree with what an ideal physics would entail about the situation.

This means that Humean physical patterns, such as relative frequencies and regularities about associations between event types, must be accurately predicted by a causal model that is successfully meeting the aims of causal modelling. This means at least that:

1. The pattern of probabilistic associations in a causal model must be statistically accurate to the observed frequencies (§1). This notion of accuracy doesn't require perfect match with observed frequencies, only that the theoretical probabilities should fit the frequency evidence.
2. Causal relations support counterfactuals; these must agree with the results of any experimental interventions designed to test dependencies (such as RCTs), as well as agree with any counterfactual dependencies that follow from the underlying physics under certain idealizing approximative assumptions. For example, the assumption of physical isolation or quasi-isolation allows us to extract counterfactuals from physical models, because it in effect characterizes some possible differences between models as gratuitous, and facilitates our focus on a subclass of models that then fix what would happen under a certain intervention (§3). So we certainly want any acceptable causal model to be true in what it says about isolated local systems.

Those are empirical constraints on causal models, helping us decide which models we might accept. We also want to evaluate their outcomes, in the sense that a good causal model should facilitate effective choice, at least in the long run, by an agent's own lights. So any successful causal model M should be such that an agent whose credences are well regulated—either well calibrated (van Fraassen 1983) or well matched to their expectation of the chances (Lewis 1986)—and who accepts a causal partition based on M, should in the long run not do systematically worse than they would have if they had relied on some other candidate causal model.

That is, adequate causal models must be among the best available to the agent in terms of serving their deliberative ends. This conception of adequacy means that an adequate causal model can come to be inadequate if better rivals are constructed, so perhaps we might say that a model is inadequate if there is some extant rival such that in the long run those who use the rival to set their credences tend to achieve valuable outcomes (according to them).

9. The Fictionalist Attitude

I have already indicated that fictionalists need not take an attitude of 'making believe' or pretence to the content of the fiction (§6). Given what I've said, especially about the role causal models play in practice, we need an attitude that is not wholly non-committal, but which is such that taking that attitude to different models could lead to different actions.

Van Fraassen suggests an attitude he calls *acceptance* as the appropriate one to take to scientific theories:

> to accept a theory is (for us) to believe that it is empirically adequate—that what the theory says *about what is observable* (by us) is true.
>
> (van Fraassen 1980, 18)

This is the cognitive component of acceptance; to accept P just is to believe some weaker proposition Ap. But to accept a theory is linked with practice in a way that belief in the weaker proposition is not, because it is also associated with a practical commitment to deploy the theory in scientific reasoning (rather than to deploy only the claim that things are empirically just as if it were true). It involves 'a commitment to confront any future phenomena by means of the conceptual resources of this theory' (van Fraassen 1980, 12).

Of course, causal fictionalists would be well advised not to use van Fraassen's notion, because many causal models with merely unobservable differences might nevertheless lead to different counterfactual predictions, and hence different recommendations in action. But there is an obvious candidate attitude in the vicinity, given what we've said about the aims of causal science:

Guidance   To be *guided* by a theory is to believe that it is physically adequate, and to undertake to frame decision problems and explanations using causal and counterfactual information provided by the theory.

This commitment is implemented in a distinctive way in our discussion above. Recall that causal propositions are not directly the objects of credence in our framework. Rather, causal propositions establish the background structure, either selecting a privileged partition of the outcome space or structuring what is held fixed under counterfactual suppositions. In neither case are we invited to consider any propositions with 'causation' as a constituent. In the admissions example, the partition was over a family of propositions about the department applied to—not over propositions about how the department applied to is causally related to your admissions prospects. Likewise in explanation: the explanatory counterfactual is the one evaluated in a context in which the department applied to is held fixed because it is a causally relevant factor—but no propositions about causally relevant factors are themselves objects of credence. The notion of guidance above fits this neatly—that consequences of coming to be guided by a theory aren't exhausted by changes in one's cognitive attitudes, but also involve the choices about framing one makes, which may not appear in one's credences in any obvious way. Full belief too might lead one to the same sort of framing, but that would differ from guidance precisely on the question of belief in the theory.

To engage quasi-assertorically in causal discourse is to express your commitments. To say, 'which department you are applying to is among the contributing causes of whether you will be admitted' is, on this view, to express your commitment to a certain causal partition, and (also) a recommendation to your hearers to adopt the same commitment. According to Joyce, moral fictionalists make moral claims to bolster, at an individual and group level, the precommitments those moral claims embody. Likewise, causal fictionalists make causal claims to endorse certain structural assumptions as approvable in practical activities.

This may be contentious, as this kind of expressive function for causal language may seem to be invoking a novel speech act not elsewhere attested. However, I wonder if this is actually a by-product of too narrow a conception of assertion. If the causal fictionalist is right, there are two different kinds of attitude that both broadly endorse the content of a claim—believing/having high credence in the claim, and adopting that claim to scaffold one's beliefs/credences. If assertion is (roughly) the speech act one opts for in wanting to endorse the content of what one says, then it may be that some assertions all along were attempts to endorse their contents as apt choices to scaffold one's credal structure. This idea would require further development, but I'm sufficiently assured by it to be confident that there is a defensible interpretation of causal discourse as not committing all of its participants to overt belief in causal claims.

10. Conclusion

Fictionalist proposals are very alluring to metaphysicians. They hold out the promise of avoiding commitment to problematic entities while keeping the benefits of talking about them. After an initial flurry of interest in fictionalism about possible worlds (Rosen 1990), numbers (Yablo 2001), and morality, the discussion waned somewhat. No doubt this is partly to do with philosophical sociology and fashion—perhaps the rise of grounding? (Which has itself recently been offered a fictionalist treatment: Thompson (2021).) But it is also due to the difficulty of spelling out decent grounds for scepticism about *F*s while making a powerful case that *F* talk yields irreplaceable benefits. I think the discussion above illustrates this; the long subsection on retention (§7) was largely devoted to trying to show that non-causal explanation and decision was insufficient on its own. That section was hard but, I think, necessary work before we can reap the purported benefits of fictionalism.

One major issue tempts further inquiry. We've seen that the indispensability of causation is intimately bound up with counterfactual and probabilistic information. If we've been led to fictionalism about causation, ought we, for consistency's sake, also be fictionalists about modality? Some time ago, Stalnaker said:

> Sometimes I am tempted to believe that there is only an actual world. But we do represent to ourselves pictures of ways that things might be, or might have been, and this practice is not just the idle exercise of our imaginations; it is central to some of our more serious activities such as giving scientific explanations of how and why the *actual* world works the way it does.   (Stalnaker 1979, 354)

This might form the germ of a modal fictionalism that really deserves the name: a view on which modal distinctions themselves are artefacts of a representational stance that is pragmatically unavoidable for us. Such a project is radical enough to make causal fictionalism seem mundane, but would seem to be motivated by some of the same concerns leading us to causal fictionalism here.[18]

References

Ahmed, Arif. 2014. *Evidence, Decision and Causality*. Cambridge: Cambridge University Press.

Albert, David Z. 2013. 'Wave Function Realism.' In *The Wave Function: Essays on the Metaphysics of Quantum Mechanics*, edited by Alyssa Ney and David Z Albert, 52-7. Oxford: Oxford University Press. https://doi.org/10.1093/acprof:oso/9780199790807.003.0001.

[18] Thanks to an anonymous referee and Yafeng Shan for comments, and particularly to Yafeng for his forbearance. I wish also to acknowledge the support of the Australian Research Council under grant DP200100190.

Bickel, P. J., E. A. Hammel, and J. W. O'Connell. 1975. 'Sex Bias in Graduate Admissions: Data from Berkeley.' *Science* 187 (4175): 398–404. https://doi.org/10.1126/science.187.4175.398.

Burgess, John P., and Gideon Rosen. 1997. *A Subject with No Object*. Oxford: Oxford University Press.

Caddick Bourne, Emily. 2013. 'Fictionalism.' *Analysis* 73 (1): 147–62. https://doi.org/10.1093/analys/ans126.

Cartwright, Nancy. 1979. 'Causal Laws and Effective Strategies.' *Noûs* 13: 419–37. https://doi.org/10.2307/2215337.

Cartwright, Nancy. 1983. *How the Laws of Physics Lie*. Oxford: Clarendon Press.

Cartwright, Nancy. 1989. *Nature's Capacities and Their Measurement*. Oxford: Oxford University Press.

Cartwright, Nancy. 1999. *The Dappled World: A Study of the Boundaries of Science*. New York: Cambridge University Press.

Cartwright, Nancy, and Jeremy Hardie. 2012. *Evidence-Based Policy: A Practical Guide to Doing It Better*. Oxford: Oxford University Press.

Contessa, Gabriele. 2010. 'Scientific Models and Fictional Objects.' *Synthese* 172 (2): 215–29. https://doi.org/10.1007/s11229-009-9503-2.

Curiel, Erik. 2021. 'Singularities and Black Holes.' In *The Stanford Encyclopedia of Philosophy*, edited by Edward N. Zalta, Fall 2021. Metaphysics Research Lab, Stanford University. https://plato.stanford.edu/archives/fall2021/entries/spacetime-singularities/.

Dawes, Gregory W. 2017. 'Ancient and Medieval Empiricism.' In *The Stanford Encyclopedia of Philosophy*, edited by Edward N. Zalta, Winter 2017. Metaphysics Research Lab, Stanford University. https://plato.stanford.edu/archives/win2017/entries/empiricism-ancient-medieval/.

Demarest, Heather, and Michael Townsen Hicks. 2021. 'Isolation, Not Locality.' *Philosophy and Phenomenological Research* 103 (3): 607–19. https://doi.org/10.1111/phpr.12731.

Dorr, Cian. 2016. 'Against Counterfactual Miracles.' *Philosophical Review* 125 (2): 241–86. https://doi.org/10.1215/00318108-3453187.

Eagle, Antony. 2007a. 'Pragmatic Causation.' In *Causation, Physics and the Constitution of Reality: Russell's Republic Revisited*, edited by Huw Price and Richard Corry, 156–90. Oxford: Oxford University Press.

Eagle, Antony. 2007b. 'Telling Tales.' *Proceedings of the Aristotelian Society* 107 (2): 125–47. https://doi.org/10.1111/j.1467-9264.2007.00215.x.

Earman, John. 1986. *A Primer on Determinism*. University of Western Ontario Series in Philosophy of Science, Vol. 32. Dordrecht: D. Reidel.

Eklund, Matti. 2019. 'Fictionalism.' In *The Stanford Encyclopedia of Philosophy*, edited by Edward N. Zalta, Winter 2019. Metaphysics Research Lab, Stanford University. https://plato.stanford.edu/archives/win2019/entries/fictionalism/.

Elga, Adam. 2007. 'Isolation and Folk Physics.' In *Causation, Physics and the Constitution of Reality: Russell's Republic Revisited*, edited by Huw Price and Richard Corry, 106–19. Oxford: Oxford University Press.

Field, Hartry. 2003. 'Causation in a Physical World.' In *Oxford Handbook of Metaphysics*, edited by Michael J. Loux and Dean W. Zimmerman, 435–60. Oxford: Oxford University Press. https://doi.org/10.1093/oxfordhb/9780199284221.003.0015.

Fine, Kit. 2005. 'Tense and Reality.' In *Modality and Tense*, 261–320. Oxford: Oxford University Press.

Fisher, R. A. 1935. *The Design of Experiments*. Edinburgh: Oliver & Boyd.

Fodor, Jerry. 1974. 'Special Sciences, or the Disunity of Science as a Working Hypothesis.' *Synthese* 28 (2): 97–115. https://doi.org/10.1007/bf00485230.

Fodor, Jerry. 1997. 'Special Sciences: Still Autonomous After All These Years.' *Philosophical Perspectives* 11 (June): 149–63. https://doi.org/10.1111/0029-4624.31.s11.7.

Frigg, Roman. 2010. 'Models and Fiction.' *Synthese* 172 (2): 251–68. https://doi.org/10.1007/s11229-009-9505-0.

Frisch, Mathias. 2014. *Causal Reasoning in Physics*. Cambridge: Cambridge University Press.

Gibbard, Allan, and William L. Harper. 1978. 'Counterfactuals and Two Kinds of Expected Utility.' In *Foundations and Applications of Decision Theory*, edited by C. A. Hooker, J. J. Leach, and E. F. McClennan, 1: 125–62. Dordrecht: D. Reidel.

Godfrey-Smith, Peter. 2009. 'Models and Fictions in Science.' *Philosophical Studies* 143 (1): 101–16. https://doi.org/10.1007/s11098-008-9313-2.

Granger, C. W. J. 1969. 'Investigating Causal Relations by Econometric Models and Cross-Spectral Methods.' *Econometrica* 37 (3): 424–38. https://doi.org/10.2307/1912791.

Hesslow, Germund. 1976. 'Two Notes on the Probabilistic Approach to Causality.' *Philosophy of Science* 43: 290–2. http://www.jstor.org/stable/187270.

Hill, A. Bradford. 1965. 'The Environment and Disease: Association or Causation?' *Proceedings of the Royal Society of Medicine* 58 (May): 295–300. https://doi.org/10.1177/003591576505800503.

Hitchcock, Christopher. 2001. 'The Intransitivity of Causation Revealed in Equations and Graphs.' *Journal of Philosophy* 98 (6): 273–99. http://www.jstor.org/stable/2678432.

Hitchcock, Christopher. 2007. 'What Russell Got Right.' In *Causation, Physics and the Constitution of Reality: Russell's Republic Revisited*, edited by Huw Price and Richard Corry, 45–65. Oxford: Oxford University Press.

Hitchcock, Christopher. 2013. 'What Is the "Cause" in Causal Decision Theory?' *Erkenntnis* 78 (S1): 129–46. https://doi.org/10.1007/s10670-013-9440-9.

Hoover, Kevin D. 2008. 'Causality in Economics and Econometrics.' In *The New Palgrave Dictionary of Economics*, edited by Steven N. Durlauf and Lawrence E. Blume, 1–13. London: Palgrave Macmillan. https://doi.org/10.1057/978-1-349-95121-5_2227-1.

Howick, Jeremy, Iain Chalmers, Paul Glasziou, Trish Greenhalgh, Carl Heneghan, Alessandro Liberati, Ivan Moschetti, et al. 2011. 'The Oxford 2011 Levels of Evidence.' Oxford Centre for Evidence-Based Medicine. http://www.cebm.net/index.aspx?o=5653.

Ismael, Jenann. 2015. 'How Do Causes Depend on Us? The Many Faces of Perspectivalism.' *Synthese* 193 (1): 245–67. https://doi.org/10.1007/s11229-015-0757-6.

Joyce, James M. 1999. *The Foundations of Causal Decision Theory*. Cambridge: Cambridge University Press.

Joyce, Richard. 2005. 'Moral Fictionalism.' In *Fictionalism in Metaphysics,* edited by Mark Eli Kalderon, 287–313. Oxford: Oxford University Press.

Kalderon, Mark Eli. 2005. *Moral Fictionalism*. Oxford: Oxford University Press.

Kim, Seahwa. 2005. 'Modal Fictionalism and Analysis.' In *Fictionalism in Metaphysics*, edited by Mark Eli Kalderon, 116–33. Oxford: Oxford University Press.

Kroon, Frederick. 2011. 'Fictionalism in Metaphysics.' *Philosophy Compass* 6 (11): 786–803. https://doi.org/10.1111/j.1747-9991.2011.00442.x.

Latham, Noa. 1987. 'Singular Causal Statements and Strict Deterministic Laws.' *Pacific Philosophical Quarterly* 68 (1): 29–43.

Levitz, Lauren, Mark Janko, Kashamuka Mwandagalirwa, Kyaw L. Thwai, Joris L. Likwela, Antoinette K. Tshefu, Michael Emch, and Steven R. Meshnick. 2018. 'Effect of Individual and Community-Level Bed Net Usage on Malaria Prevalence among Under-Fives in the Democratic Republic of Congo.' *Malaria Journal* 17 (1). https://doi.org/10.1186/s12936-018-2183-y.

Lewis, David. 1973. *Counterfactuals*. Oxford: Blackwell.

Lewis, David. 1981. 'Causal Decision Theory.' *Australasian Journal of Philosophy* 59 (1): 5–30. https://doi.org/10.1080/00048408112340011.

Lewis, David. 1986. 'A Subjectivist's Guide to Objective Chance.' In *Philosophical Papers*, 2: 83–132. Oxford: Oxford University Press.

Lewis, David. 2005. 'Quasi-Realism Is Fictionalism.' In *Fictionalism in Metaphysics*, edited by Mark Eli Kalderon, 314–21. Oxford: Oxford University Press.

Lindstrom, Lamont. 1993. *Cargo Cult*. Honolulu: University of Hawai'i Press. https://doi.org/10.2307/j.ctv9zcktq.

Loewer, Barry. 2009. 'Why Is There Anything Except Physics?' *Synthese* 170 (2): 217–33. https://doi.org/10.1007/s11229-009-9580-2.

Lucas, Robyn M., and Anthony J. McMichael. 2005. 'Association or Causation: Evaluating Links between "Environment and Disease".' *Bulletin of the World Health Organization* 83 (10): 792–5.

Maudlin, Tim. 2021. 'Relativity and Space-Time Geometry.' In *The Routledge Companion to Philosophy of Physics*, edited by Eleanor Knox and Alastair Wilson, 61–70. New York and London: Routledge.

Menzies, Peter. 2007. 'Causation in Context.' In *Causation, Physics and the Constitution of Reality: Russell's Republic Revisited*, edited by Huw Price and Richard Corry, 191–223. Oxford: Oxford University Press.

Menzies, Peter, and Christian List. 2010. 'The Causal Autonomy of the Special Sciences.' In *Emergence in Mind*, edited by Cynthia McDonald and Graham McDonald, 108–29. Oxford: Oxford University Press.

Mill, John Stuart. (1874) 1974. *A System of Logic, Ratiocinative and Inductive, Books I–III*. Edited by John M. Robson. 8th ed. The Collected Works of John Stuart Mill. Toronto: University of Toronto Press; London: Routledge & Kegan Paul.

Nagel, Thomas. 1986. *The View from Nowhere*. Oxford: Oxford University Press.

Ney, Alyssa. 2015. 'Fundamental Physical Ontologies and the Constraint of Empirical Coherence: A Defense of Wave Function Realism.' *Synthese* 192 (10): 3105–24. https://doi.org/10.1007/s11229-014-0633-9.

Norton, John D. 2003. 'Causation as Folk Science.' *Philosophers' Imprint* 3 (4): 1–22. http://hdl.handle.net/2027/spo.3521354.0003.004.

Paul, L. A., and Ned Hall. 2013. *Causation: A User's Guide*. Oxford: Oxford University Press. https://doi.org/10.1093/acprof:oso/9780199673445.001.0001.

Pearl, Judea. 2000. *Causality: Models, Reasoning and Inference*. Cambridge: Cambridge University Press.

Peterson, Martin. 2017. *An Introduction to Decision Theory*. 2nd ed. Cambridge: Cambridge University Press. https://doi.org/10.1017/9781316585061.

Potts, Christopher. 2015. 'Presupposition and Implicature.' In *The Handbook of Contemporary Semantic Theory*, edited by Shalom Lappin and Chris Fox, 2nd ed., 168–202. Oxford: Wiley-Blackwell. https://doi.org/10.1002/9781118882139.ch6.

Price, Huw. 1986. 'Against Causal Decision Theory.' *Synthese* 67 (2): 195–212. https://doi.org/10.1007/bf00540068.

Price, Huw. 2007. 'Causal Perspectivalism.' In *Causation, Physics and the Constitution of Reality: Russell's Republic Revisited*, edited by Huw Price and Richard Corry, 250–92. Oxford: Oxford University Press.

Ramsey, F. P. (1929) 1990. 'General Propositions and Causality.' In *Philosophical Papers*, edited by D. H. Mellor, 145–63. Cambridge: Cambridge University Press.

Robertson, Katie. Forthcoming. 'Autonomy Generalised; or, Why Doesn't Physics Matter More?' *Ergo*.

Rosen, Gideon. 1990. 'Modal Fictionalism.' *Mind* 99: 327–54. http://www.jstor.org/stable/2255102.

Russell, Bertrand. 1913. 'On the Notion of Cause.' *Proceedings of the Aristotelian Society* 13 (1): 1–26. https://doi.org/10.1093/aristotelian/13.1.1.

Sainsbury, R. M. 2010. *Fiction and Fictionalism*. London and New York: Routledge.

Schaffer, Jonathan. 2010. 'Causation, Physics, and the Constitution of Reality: Russell's Republic Revisited, Edited by Huw Price and Richard Corry.' *Mind* 119 (475): 844–8. https://doi.org/10.1093/mind/fzq068.

Simon, Herbert A. 1977. 'Causal Ordering and Identifiability.' In *Models of Discovery*, 53–80. Boston Studies in the Philosophy of Science, Vol. 54. Dordrecht: Springer Netherlands. https://doi.org/10.1007/978-94-010-9521-1_5.

Skyrms, Brian. 1980. *Causal Necessity*. New Haven, CT: Yale University Press.

Solomonoff, Ray. 1964. 'A Formal Theory of Inductive Inference, Part I.' *Information and Control* 7: 1–22. https://doi.org/10.1016/S0019-9958(64)90223-2.

Spirtes, Peter, Clark Glymour, and Richard Scheines. 2000. *Causation, Prediction and Search*. Cambridge, MA: MIT Press.

Stalnaker, Robert. 1979. 'Anti-essentialism.' *Midwest Studies in Philosophy* 4 (1): 343–55. https://doi.org/10.1111/j.1475-4975.1979.tb00385.x.

Stanley, Jason. 2001. 'Hermeneutic Fictionalism.' *Midwest Studies in Philosophy* 25: 36–71. http://www.blackwell-synergy.com/doi/abs/10.1111/1475-4975.00039.

Stovitz, Steven D., and Ian Shrier. 2019. 'Causal Inference for Clinicians.' *BMJ Evidence-Based Medicine* 24 (3): 109–12. https://doi.org/10.1136/bmjebm-2018-111069.

Suppes, Patrick. 1970. *A Probabilistic Theory of Causality*. Amsterdam: North-Holland.

Thompson, Naomi. 2021. 'Setting the Story Straight: Fictionalism about Grounding.' *Philosophical Studies* 179: 343–61. https://doi.org/10.1007/s11098-021-01661-w.

van Fraassen, Bas C. 1980. *The Scientific Image*. Oxford: Clarendon Press. https://doi.org/10.1093/0198244274.001.0001.

van Fraassen, Bas C. 1983. 'Calibration: A Frequency Justification for Personal Probability.' In *Physics, Philosophy and Psychoanalysis*, edited by Robert S. Cohen and Larry Laudan, 295–319. Boston Studies in the Philosophy of Science, Vol. 76. Dordrecht: D. Reidel.

van't Hoff, Alice. 2022. 'In Defense of Causal Eliminativism.' *Synthese* 200 (393). https://doi.org/10.1007/s11229-022-03875-9.

Wallace, David, and C. G. Timpson. 2010. 'Quantum Mechanics on Spacetime I: Spacetime State Realism.' *British Journal for the Philosophy of Science* 61 (4): 697–727. https://doi.org/10.1093/bjps/axq010.

Williamson, Jon. 2019. 'Establishing Causal Claims in Medicine.' *International Studies in the Philosophy of Science* 32 (1): 33–61. https://doi.org/10.1080/02698595.2019.1630927.

Woodward, James. 2003. *Making Things Happen*. Oxford: Oxford University Press.

Woodward, Jim. 2002. 'What Is a Mechanism? A Counterfactual Account.' *Philosophy of Science* 69 (S3): S366–77. https://doi.org/10.1086/341859.

Yablo, Stephen. 2001. 'Go Figure: A Path through Fictionalism.' *Midwest Studies in Philosophy* 25 (1): 72–102. https://doi.org/10.1111/1475-4975.00040.

Zimmerman, Dean W. 2008. 'The Privileged Present: Defending an "A-theory" of Time.' In *Contemporary Debates in Metaphysics*, edited by Theodore Sider, John Hawthorne, and Dean W. Zimmerman, 211–25. Oxford: Blackwell.

10

Epistemic Causality and its Application to the Social and Cognitive Sciences

Yafeng Shan, Samuel D. Taylor, and Jon Williamson

1. Introduction

Just as the epistemic theory of probability (i.e. Bayesianism) interprets probability as a kind of belief—namely, rational probabilistic belief—the epistemic theory of causality interprets causality as a kind of belief—rational causal belief, in this case. The epistemic theory was put forward by Williamson (2005, ch. 9) and further discussed and developed by Williamson (2006a, 2006b, 2007, 2009, 2011, 2013, 2021), Russo (2009, §7.5), Russo and Williamson (2007, 2011a, 2011b), Wilde and Williamson (2016), and Taylor (2021). This chapter provides a brief introduction to epistemic causality in §2, and then motivates the theory from a general point of view and by examining scientific practice.

In §3 we sketch four arguments for epistemic causality: an argument from the failure of more standard approaches (the argument from *failure*, §3.1), an argument that stems from the desideratum of parsimony (the argument from *parsimony*, §3.2), an argument that stems from a particular view of the epistemology of causality (the argument from *Evidential Pluralism*, §3.3), and an argument from the need to remain neutral about certain questions (the argument from *neutrality*, §3.4).

In §4 we argue that the epistemic theory provides a natural interpretation of causality in the social sciences, illustrating the arguments from failure, parsimony, and Evidential Pluralism. In §5 we show that epistemic causality is also well suited to cognitive science, by appeal to the arguments from Evidential Pluralism and neutrality. We conclude in §6 that the epistemic theory provides a philosophical account of causality that is well motivated across a range of sciences.

2. The Epistemic Theory of Causality

In this section, we provide an introduction to the epistemic theory of causality. Although the epistemic theory is a theory of the nature of causality, it is intimately connected to the epistemology of causality, as we shall now describe.

Yafeng Shan, Samuel D. Taylor, and Jon Williamson, *Epistemic Causality and its Application to the Social and Cognitive Sciences* In: *Alternative Approaches to Causation: Beyond Difference-making and Mechanism*. Edited by: Yafeng Shan, Oxford University Press. © Yafeng Shan 2024. DOI: 10.1093/oso/9780192863485.003.0010

Clearly, we have causal beliefs and make causal claims. These help us to successfully predict, explain, and control our world. We shall refer to these predictions, explanations, and control inferences that are characteristic of causal claims as 'PECs'.

Our causal beliefs and claims depend on our evidence. In some cases, these beliefs and claims are appropriate, given the evidence, while in others they are not. (Whether causal claims are appropriate given the evidence is a different question to whether they are borne out by further enquiry.) We refer to a theory that says something about which causal claims are appropriate given the available evidence E as a *causal epistemology*.

A variety of causal epistemologies have been put forward in the literature. In the literature on causal cognition, there are accounts that emphasize counterfactual reasoning (Gerstenberg et al., 2022), interventions (Gopnik et al., 2004), temporal cues (Lagnado and Sloman, 2004), mechanistic information (Ahn et al., 1995) and decision making (Sobel and Kushnir, 2006), for example. In science and medicine, there are accounts that emphasize randomized trials (Guyatt et al., 1992; Sackett et al., 1996), quantitative methods (Imai, 2017), qualitative methods (Patton, 2002; Glynn and Ichino, 2015), or mixed methods (Teddlie and Tashakkori, 2009).

This plethora of approaches highlights three points. Firstly, there are clearly many diverse indicators of causality. Second, there is no settled view as to how to capture these indicators in a single causal epistemology. Third, some causal epistemologies may be better than others. In medicine, for example, there has been a transition from an approach based on authority and experience to 'evidence-based' medicine (EBM). This transition has been accompanied by improvements to health outcomes. Proponents of the EBM+ approach to evidence-based medicine have argued that further improvements may be possible, by making additional changes to our causal epistemology (Parkkinen et al., 2018).

This third point raises the possibility that we are progressing towards some optimal causal epistemology—albeit an ideal that may never in fact be reached. What would it take for a causal epistemology to be optimal? Such a theory would need to balance several demands. Putative desiderata include the following. (i) Reliability: a causal epistemology should yield a body of causal claims that underwrites successful PECs. (ii) Strength: it should establish and rule out sufficiently many claims for science to progress efficiently. (iii) Stability: those claims that are established or ruled out should likely remain so in the face of subsequent evidence.[1] (iv) Completeness: it should determine, for any putative causal claim, whether the claim is established or ruled out by evidence, or, if neither, what

---

[1] Some of these desiderata are general epistemological desiderata, not specific just to causal epistemologies. For example, Stability is a desideratum associated with establishing in general and is not specific to establishing causal claims (Williamson, 2022). When establishing a proposition we expect to be able to use it as evidence for other propositions in the long term. It would be almost

degree of confidence is warranted in the claim. (v) Simplicity: it should not be unnecessarily complicated—for example, single-case causal claims should be subsumable under generic causal claims as far as possible. (vi) Feasibility: it should be practicable to use the causal epistemology to evaluate causal claims of interest.[2]

According to the epistemic theory of causality, these ingredients are all we need for an adequate theory of causality: *the causal facts are just facts about what is established or ruled out by all optimal causal epistemologies on total evidence.* A is a cause of B just if every optimal causal epistemology would deem the claim that A is a cause of B to be established by an idealized evidence base that consists of all matters of particular fact—past, present, and future. A is not a cause of B iff every causal epistemology deems the claim that A is a cause of B to be ruled out by total evidence. Otherwise—if some optimal causal epistemologies deem A to be a cause of B and others not—it is indeterminate whether A is a cause of B.

According to the epistemic theory, then, there is no need to take causality to be 'out there', nor analysable in terms of a single indicator of causality, such as probabilistic dependence, counterfactual connection, mechanistic connection, temporal succession, etc. The reality is that all these indicators play a role in our judgements about what causes what, and attempts to give one primacy over the others invariably fail. To understand causality we need to understand the roles these indicators play in fixing our causal beliefs. The facts of causality are facts about rational belief, not facts about some non-epistemological connection between the causal relata.

The epistemic understanding of causality in terms of rational belief is analogous to the Bayesian understanding of probability in terms of rational belief. Thus a parallel can be drawn between epistemic causality and epistemic probability. A probabilistic belief is a kind of belief, not a belief about some kind of non-epistemic probability: it is a relational belief of the form $P_E(A) = x$, which says that rational degree of belief in A on evidence E equals x. Our probabilistic beliefs enable successful predictions, decisions, and actions, and the facts about rational probabilistic belief are determined by optimal probabilistic epistemologies, not by a single indicator of probability, such as symmetry, observed frequency, or confirmed theory. Similarly, causal belief is a kind of belief, not a belief about some kind of non-epistemic causal relation: it is a relational belief of the form $C_E(A,B)$, and a body of such beliefs enable characteristic PECs. The analogy between epistemic causality and epistemic probability can be pushed further: Williamson (2021) argues that the two approaches admit analogous norms and yield analogous analyses of the relevant facts.

Moore-paradoxical to say that 'We have established that smoking causes cancer but we expect to retract this claim in the coming year.' While some established claims will inevitably be overturned in the light of new evidence, the likelihood of this happening for any given claim ought to be small.

[2] The difference between Completeness and Feasibility is that the latter requires that the epistemology be one that could be applied in practice, while the former does not.

3. The Case for Epistemic Causality

Having introduced the key tenets of epistemic causality, we shall now briefly sketch four ways of motivating this view of causality: by appeal to the inadequacy of alternative accounts (§3.1), by appeal to parsimony (§3.2), by appeal to a recent view of the epistemology of causality (§3.3), or by appeal to the need to remain neutral about certain questions (§3.4). These arguments will be explored in more detail in subsequent sections.

3.1 The Argument from Failure

The argument from failure proceeds from the observation that other theories of causality tend to fall to counterexamples, while epistemic causality does not. These considerations favour the latter theory over the former theories.

Standard theories of causality are often classed as difference-making or mechanistic theories. Difference-making theories include the regularity, counterfactual, probabilistic, agency, and interventionist theories. All these theories require that a cause should make the appropriate kind of difference to its effects. Unfortunately, one can usually find cases of causation where the cause cannot make a difference, because the effect is already fully determined by other factors (see, e.g., Hall, 2004). One strategy here is to suggest that the cause would made a difference when the other factors are absent, but even this strategy does not work when the set of causes is mutually exclusive and exhaustive, because then the cause of interest cannot be varied independently of the others (Williamson, 2005, §7.3). Mechanistic theories, on the other hand, require that cause and effect should be connected by some appropriate kind of mechanism. These theories face problems in cases where the cause or the effect is an absence of something, since an absence cannot be a part of a mechanism (see, e.g., Hall, 2004; Williamson, 2011, §II.1). The behaviour of a mechanism is supposed to be explained by the arrangement of its (actually present) constituents, such as entities, activities, and events.

A common suggestion is to move to pluralism. One approach here is a kind of dualism: the view that some causal claims are claims about difference-making while other claims are claims about mechanisms (Hall, 2004). This view falls to counterexamples in which there is neither difference-making nor a linking mechanism (Longworth, 2006). Moreover, our use of causal talk stands at odds with pluralism. We do not tend to ask clarifying questions to disambiguate a causal claim, in the way that we might with some probabilistic claim which could be interpreted either as a claim about rational degree of belief or as a claim about frequency. This latter problem also besets more radical kinds of pluralism, such as the inferentialism of Reiss (2012)—see Williamson (2006a, 2013) for further discussion.

Epistemic causality does not succumb to these problems that beset pluralism because it is a monistic theory: although there are multiple indicators of causality (evidential pluralism), there is a single concept of cause (conceptual monism) and a single causal relation (metaphysical monism). Epistemic causality is metaphysically monist because it analyses the causal relation solely in terms of facts about what is established or ruled out by all optimal causal epistemologies on total evidence. There is no other causal relation, according to epistemic causality, and this is evidenced by the decisive objections that face other accounts of causality. Epistemic causality is conceptually monist because it invokes a single rational-belief concept of cause. Again, no other concept of cause is viable, according to epistemic causality, as is witnessed by objections to other concepts of cause and to pluralism.

Additionally, epistemic causality does not succumb to the problem of counter-examples involving cases of overdetermination or absences. Indeed it is hard to see how one could produce any sort of counterexample to epistemic causality. A counterexample to epistemic causality would require finding a causal fact that the epistemic theory misclassifies. Recall that the epistemic theory deems A to be a cause of B just when every optimal causal epistemology deems that A is a cause of B, on total evidence. It deems A not to be a cause of B iff every optimal causal epistemology deems that A isn't a cause of B. It is indeterminate whether A is a cause of B iff some optimal causal epistemology says it is and some other says it isn't. A counterexample to the epistemic theory would need to show that the theory misclassifies some particular causal claim, and it would need to do this by appeal to some consideration that settles the question of the correct classification. The difficulty is that if this consideration were correct and conclusive then one would expect that it would be validated by each optimal causal epistemology. For instance, if the example were one of causation between absences, and our intuitions about the example are correct, then one would anticipate that every optimal causal epistemology would validate the example. So it is hard to see how any counterexample put forward against the epistemic theory could be conclusive.

Thus, the problems that beset the standard accounts of causality favour the epistemic theory over these rival accounts.

3.2 The Argument from Parsimony

A second line of argument for the epistemic theory appeals to the idea of parsimony, as follows. Clearly we have causal beliefs and make causal claims and we need to theorize about how best to do this, in order to progress science, medicine, public policy, and our own decision making. So we need causal epistemologies and we need to think about how good they are. Then it is but a small step to epistemic causality: its building blocks are just causal epistemologies and the idea of progress towards an ideal causal epistemology. Moreover, the epistemic theory is successful in yielding the correct judgements of causality, as we saw above when we considered the possibility of counterexamples to epistemic causality. Given that the epistemic theory is itself parsimonious and successful, it would be otiose to posit any additional kind of causality. There is simply no need for a further theory that attempts to analyse causality in terms of one of its indicators, or in terms of a pluralist panoply of indicators. Epistemic causality is all we need.

Let us consider three potential responses to this argument from parsimony.

Firstly, one might ask whether some alternative approach to causality can run the same sort of argument. If we analysed causality in terms of some nonepistemic X, perhaps X-causality would be all we need and it would be otiose to consider epistemic causality. There is an asymmetry here, however. The proponent of epistemic causality is likely to find X-causality to be less parsimonious: the alternative theory of causality requires some appropriate stuff 'out there' to which causality can be ultimately reduced. (This might be possible worlds, required to underwrite modal difference-making claims, or causal powers, for example.) On the other hand, even an advocate of X-causality should admit that we need to consider causal beliefs and causal epistemologies, whether or not causality is analysable in terms of X.

Second, one might think that even if alternative theories of causality are otiose as analyses of causality, they may yet have some heuristic value. For example, a counterfactual theory of causation can suggest certain strategies for testing causation—strategies which may have led to some improvements to past causal epistemologies. But note that one can admit this heuristic role for alternative theories of causality while taking epistemic causality to be the correct analysis of causality. Thus there is no incompatibility here. On the other hand, the heuristic value of alternative theories is by no means clear cut: identifying causality with counterfactual connection may have hampered more than it has helped, for example. Indeed, proponents of EBM+ might argue that this identification has merely delayed potential improvements to causal epistemology in medicine, by entrenching the position of present-day EBM, which has important limitations.[3]

Third, one might worry that we have a slippery slope: that if we accepted that parsimony motivates epistemic causality then we would be forced to accept that parsimony motivates an epistemic theory of everything. But we would not want to be forced to adopt an epistemic theory of tables and chairs, for example. The concern is that this would ultimately be too revisionary.

---

[3] The tendency to conflate causality with some specific indicator of causality, such as counterfactual connection, can be viewed as an instance of what Jaynes (2003) called the 'mind projection fallacy'—the mistake made by construing something that is really epistemic to be a feature of the non-epistemic world. Its refusal to project causal relationships onto the world sets epistemic causality apart from projectivist theories of causality (Beebee, 2015).

One can resist an epistemic theory of everything, however, by appealing to simplicity and success. It's much simpler, given the way we think, to construe facts about tables and chairs as facts about things out there rather than as facts about rational table-beliefs and rational chair-beliefs. Moreover, taking tables and chairs to be out there is successful—not prone to counterexamples. Hence there is no need to resort to an epistemic theory of tables and chairs.

Thus, epistemic theories differ from certain global approaches—such as pragmatism, Humean supervenience, and inferentialism—in being appropriate only in those cases in which simpler theories cannot be successfully applied.

3.3 The Argument from Evidential Pluralism

A third argument for epistemic causality appeals to one particular causal epistemology; namely, Evidential Pluralism. This epistemological theory provides a very general account of the confirmation relationships involved in assessing causal claims—general enough to accommodate many of the indicators of causality introduced in §2. On the one hand, this causal epistemology appears to pose a serious challenge to standard accounts of causality. On the other hand, epistemic causality can accommodate Evidential Pluralism perfectly well. Hence, Evidential Pluralism favours epistemic causality over rival accounts.

Evidential Pluralism provides an account of what one needs to establish in order to establish that A is a cause of B, and an account of what sort of studies one needs to consider in order to assess causality. Figure 10.1 illustrates the main claims of Evidential Pluralism.

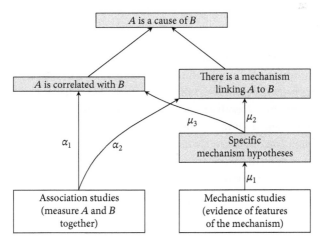

Figure 10.1 Evidential relationships for establishing a causal claim, according to Evidential Pluralism

Consider first the top part of Figure 10.1. Evidential Pluralism is motivated by the platitude that there is more to causation than correlation. What else is needed to establish a causal claim, other than evidence of a correlation conditional on potential confounders? Arguably, a correlation between A and B is attributable to causation just where there is some complex of mechanisms linking A to B according to which instances of A are partly responsible for instances of B. This motivates the thesis that establishing causality requires establishing both the existence of a correlation and the existence of an appropriate mechanistic connection (Russo and Williamson, 2007).

Moving down Figure 10.1, we turn to the question of how to establish correlation and mechanism. The usual way to establish correlation is to perform a study that repeatedly measures A and B to test for a correlation (confirmation channel α_1). Certain kinds of these association studies can also indirectly confirm the existence of a linking mechanism (α_2). In particular, high-quality randomized controlled trials can confirm the presence of a linking mechanism, by making it less likely that an observed correlation is attributable to unforeseen confounding. But there is a more direct way to establish the existence of a suitable mechanism: hypothesize key features of the mechanism and perform studies that test for the presence of these features (μ_1 and μ_2). In certain cases—especially if the details of the mechanism are well established and the mechanism itself is not too complex— this mechanistic evidence can also make the existence of a correlation more plausible (μ_3).

Evidential Pluralism poses a challenge for standard accounts of causality (Russo and Williamson, 2007). If a difference-making theory of causality were correct, it should be sufficient to establish the appropriate sort of correlation in order to establish causation: there should be no need to go on to establish the existence of a mechanism. A similar point holds for any mechanistic account of causality: if such a theory were correct, there should be no need to establish correlation once the appropriate sort of mechanism were established. Dualist accounts face the same problem. For a dualist account, some claims are difference-making claims, while others are mechanistic claims. For those that are difference-making, it should be sufficient to establish correlation. For those that are mechanistic, it should be sufficient to establish mechanism. Standard theories, then, need to either refute Evidential Pluralism or provide an account of how to accommodate Evidential Pluralism. To date, this challenge has not been adequately met.[4]

On the other hand, it is straightforward for the epistemic theory of causality to accommodate Evidential Pluralism. Indeed, a key feature of the epistemic theory is its flexibility to cope with multiple indicators of causality. If any ideal causal

---

[4] Though see Weber (2009) for an account of how Giere's probabilistic theory of causality might meet this challenge.

epistemology validates Evidential Pluralism, then, by construction, epistemic causality also validates Evidential Pluralism. Thus Evidential Pluralism favours epistemic causality over a range of standard rivals.

3.4 The Argument from Neutrality

Certain contentious positions in science are allied with either a difference-making account of causality or a mechanistic account. In order not to prejudge these contentious questions, it is important not to side with some such account. Epistemic causality does not presuppose that all causal relationships are difference-making relationships, nor that all causal relationships are mechanistic relationships. Thus the need for neutrality can favour the epistemic theory over standard alternative theories of causality.

For example, Taylor (2021) argues that epistemic causality is needed to provide a unified conception of causal explanation in cognitive science, which admits the possibility of both mechanistic and non-mechanistic explanations of cognition. The idea is that epistemic causality is required so as not to prejudge the question of whether all causal explanations in cognitive science are mechanistic: a mechanistic theory of causality would say yes, while a difference-making or dualist approach would say no. Since this question has not been settled in cognitive science, we are not in a position to ascertain whether ideal causal epistemologies will have a role for non-mechanistic causal explanations. Hence, the epistemic theory remains neutral on this question. Arguably, one needs a theory of causality that remains neutral on this question, precisely because the question hasn't been settled. This favours the epistemic theory over standard rivals.

We will consider another instance of this argument, as well as an instance of the argument from Evidential Pluralism, when we consider the cognitive sciences in §5. Meanwhile, in the next section, we will see how some of these arguments play out in the social sciences.

4. Epistemic Causality in the Social Sciences

In this section, we argue that the epistemic theory provides an account of causality that is well suited to the social sciences. We appeal here to the arguments from failure, parsimony, and Evidential Pluralism.

There are two standard approaches to causality in the social sciences: monism and pluralism (cf. Gerring, 2005; Haggard and Kaufman, 2016). First, in §4.1, we shall argue that the epistemic theory is preferable to the standard monistic approach by appeal to the argument from failure. Next, in §4.2, we shall show that the epistemic theory is preferable to the standard pluralist approach by appeal to the argument from parsimony. Finally, in §4.3, we shall argue for the epistemic theory by appeal to the argument from Evidential Pluralism.

4.1 Causal Monism in the Social Sciences and the Argument from Failure

The standard monistic approach in the social sciences typically understands causality in a difference-making way (e.g. Granger, 1969, 1980; King et al., 1994; Morgan and Winship, 2015).[5] For example, King et al. (1994, pp. 81–2) define causality in terms of 'the difference between the systematic component of observations made when the explanatory variable takes one value and the systematic component of comparable observations when the explanatory variable takes on another value'.[6] While some authors adopt a probabilistic account of difference-making, others appeal to counterfactuals (e.g. Cook and Campbell, 1979; Holland, 1986): they take causes to be factors that are manipulatable in experiments.

These difference-making approaches face serious objections. Typical objections to the probabilistic account stem from the truism that correlation is not causation. For example, it has been shown that there is a statistically significant correlation between unemployment rates and property crime in the 1990s in the United States (Gould et al., 2002; Machin and Meghir, 2004). A typical estimate would be that a 1 percentage point increase in the unemployment rate is associated with a 1 per cent increase in property crime. Based on these estimates, the observed 2 percentage point decline in the US unemployment rate between 1991 and 2001 has been taken to cause the estimated 2 per cent decline in property crime. However, as Levitt (2004) indicates, the correlation between unemployment and crime rates in this case is historically contingent. As a counterexample, the 1960s is a decade of strong economic growth with a sharp increase in crime rates. Instead, Donohue and Levitt (2001) argue that the drop in crime rates in the 1990s in the United States was caused by the legalization of abortion in the 1970s. Thus, as has been well recognized, the probabilistic variant of the monistic approach is vulnerable to the problem that establishing correlation is not sufficient for establishing causation.

The counterfactual account can also be challenged. There is a worry concerning the assumption of the manipulability of causal variables. According to some counterfactual accounts, all causal factors must be experimentally manipulable.[7]

---

[5] Social scientists who embrace difference-making theories do not necessarily have a strong views about the ontology of causation. For example, those who advocate a probabilistic account of causation do not necessarily hold views about what constitutes this probabilistic dependence.

[6] Although King et al. (1994, pp. 86–7) recognize the significance of mechanisms in causal analysis, they contend that the concept of causality is essentially difference-making rather than mechanistic.

[7] This applies to counterfactual accounts of causation (e.g. Rubin, 1974; Holland, 1986) which originated in the agricultural sciences, but arguably not to Lewis' counterfactual theory (Lewis, 1973) nor Woodward's manipulation theory (Woodward, 2003).

In other words, social scientists should not assert any causal claim about non-manipulable variables. As Goldthorpe (2001, p. 6) illustrates, 'one could discuss the association that exists between sex or race, on the one hand, and say, educational attainment, on the other. But it would be no more meaningful to speak of sex or race as being causes of such attainment than it would be to make statements about what level of education Ms M would have achieved had she been a man or Mr N had he been a woman.' If that is right, the scope of causal claims in the social sciences is much more limited than is commonly thought. In addition, there is a concern about the distinctive nature of the response of the units in experiments in the social sciences. The counterfactual account of causation was originally introduced to the social sciences, especially sociology, following its application to medical and agricultural science (Cook and Campbell, 1979). However, this approach to causality in applied natural science cannot be straightforwardly transported to a sociological context. In principle, the counterfactual account allows conceptual space for human action only in the role of experimenter or intervener. For example, in an experiment to test a fertilizer, the experimental set-up is the only source of intervention. Once the experiment is carried out, all else has to follow in the manner of plants responding to the fertilizer. But in the social sciences, the response of the units in experiments cannot be simply assumed to have the same nature as that of the units in experiments in applied natural and agricultural sciences. Consider a case of the introduction of positive discrimination in education, with the aim of reducing class or ethnic differentials in achievement. It is likely that members of those classes or ethnic groups whose children would not benefit and who might lose their competitive advantage in schools could respond in order to preserve the advantage. In this case, one crucial requirement of experimental design would be breached: the response of a unit should not be influenced by whether other units are treated or not.

The epistemic theory does not succumb to these problems, because it does not analyse the causal relation in terms of difference-making and it does not impose strict conditions on what can count as causal relata. Thus, the counterexamples to the above monist accounts cannot pose a challenge to epistemic causality. These problems can be viewed as illustrating the argument for epistemic causality from the failure of standard alternatives.

Probabilistic and counterfactual accounts of causation are not the only monistic approaches that are relevant to the social sciences. Political science and sociology might seem to presuppose a mechanistic account of causality (Beach and Pedersen, 2013; Hedström and Ylikoski, 2010). For example, the use of process tracing, case studies, and certain small-N studies might seem to establish causation just by establishing mechanism. Just as causation is not correlation, however, so causation is not easily reducible to mechanistic connection. Indeed, as Shan and Williamson (2023) argue, while specifying a mechanism can provide a narrative explanation, it falls short of establishing a causal claim. For example, political scientists might detail the social actors and their activities of a mechanism from the assassination of Archduke Ferdinand to the outbreak of World War I, but this does not establish that the assassination of Archduke Ferdinand caused the outbreak of World War I, because the war might have happened anyway.[8]

It is evident that this problem does not undermine the epistemic theory, because the epistemic theory does not define causality in terms of mechanism. The epistemic theory is thus immune to key problems that beset monistic accounts of causality because it does not conflate causation with any particular indicator of causation.

4.2 Causal Pluralism in the Social Sciences and the Arguments from Parsimony

Pluralists maintain that there are multiple concepts of causality in the social sciences (e.g. Reiss, 2009; Goertz and Mahoney, 2012; Maziarz, 2020; Rohlfing and Zuber, 2021). For example, Goertz and Mahoney (2012) argue that there are two concepts of causality in the social sciences that underlie the two key approaches to social scientific practice—namely, the quantitative and qualitative approaches. Goertz and Mahoney's argument rests on their own interpretation of Hume's theory of causality. Hume's definition of cause is as follows:

> [W]e may define a cause to be an object, followed by another, and where all the objects, similar to the first, are followed by objects similar to the second. Or, in other words, where, if the first object had not been, the second never had existed.
>
> (Hume, 1748, §7.2.29)

Traditionally, the first part of this definition is regarded as the original formulation of the regularity theory of causality, while the second part is viewed as a precursor to the counterfactual theory of causality. Goertz and Mahoney (2012, p. 76) follow this received view by calling the first part 'the constant conjunction definition' and the second part 'the counterfactual definition'. However, Goertz and Mahoney argue that these two definitions correspond to the respective views of causation of the quantitative and qualitative traditions in the social sciences:

> Hume's famous quotation contains two definitions of causation. Definition 1 suggests a constant conjunction between cause and effect, such that effects always follow causes. This definition assumes many cases and has affinities with quantitative views on causation. Definition 2 suggests a counterfactual view of causation, in which the absence of a cause leads to the absence of an outcome. This

---

[8] For more in-depth discussion of the distinction between narrative and causal explanation, see Shan and Williamson (2023, pp. 126–7).

definition is built around a single case and has important linkages to qualitative views of causation.   (Goertz and Mahoney, 2012, p. 81)

For Goertz and Mahoney, the constant conjunction definition (or definition 1) fits quantitative methods, which presume a statistical approach to establishing causal claims, while the counterfactual definition (or definition 2) fits the methods of the qualitative approach better.[9]

As Gerring (2005) indicates, the pluralist approach overstates the ontological, epistemological, and conceptual differences between causal analyses in the social sciences. Let us illustrate this problem with a famous example of sociological research: the study of socioeconomic status and health status (House et al., 1994; Link and Phelan, 1995; Adler and Newman, 2002; Pampel et al., 2010; Phelan et al., 2010). It has been shown that there is a strong association between socio-economic status and health status. For example, lower socioeconomic status is associated with the 14 major causes of death in the International Classification of Diseases (Illsley and Mullen, 1985). In addition, lower socioeconomic status is shown to be associated with lower life expectancy, higher overall mortality rates, and higher rates of infant and perinatal mortality (Dutton, 1986; Adler et al., 1994; Bosworth, 2018). However, it is debatable whether socioeconomic status is a cause of health status. Sceptics typically argue that socioeconomic status is a placeholder variable for real causes of disease that have not yet been identified. Thus Rothman (1986, p. 90) suggests that socioeconomic status is 'a correlate of many causes of diseases'.

Even for some social scientists who argue for the causal relationship between socioeconomic status and health, a strong and pervasive association between socioeconomic status and health merely provides 'a description of the social patterning of disease' (Link and Phelan, 1995, p. 82). It is widely accepted that in order to establish the causal claim that socioeconomic status is a cause of disease, one has to establish the existence of some mechanism as well as a correlation (House et al., 1994; Phelan et al., 2004). As Link and Phelan (1995, p. 82) suggest, it is necessary to identify 'the direction of causation between social conditions and health and the mechanisms that explain observed associations' for the purpose of 'establishing a causal role for social factors'.

With their collaborators, Link and Phelan have identified a variety of mechanisms linking socioeconomic status to health status (Link and Phelan, 1995; Phelan et al., 2004, 2010). It is shown that people of higher socioeconomic status possess a wide range of resources, including money, knowledge, power, and beneficial social

---

[9] It should be noted that Goertz and Mahoney (2012, pp. 81–2) also indicate that some qualitative researchers, especially those who use qualitative comparative analysis, 'may gravitate' towards definition 1.

connections, which shape health-enhancing behaviours (such as getting flu jabs, eating fruits and vegetables, and exercising regularly) and access to broad contexts that are associated with risk and protective factors of health. For example, those who have lower-status jobs more commonly have 'job strain' (i.e. a combination of high job demands and low decision latitude), which is associated with coronary heart disease (Schnall et al., 1990); people with lower socioeconomic status are more likely to smoke and be overweight, which lead to various health problems (Lantz et al., 1998); and those with lower socioeconomic status experience greater residential crowding and noise, which is linked to poorer long-term memory and reading deficits (Evans and Saegert, 2000).

Moreover, Phelan and Link argue that although there are various mechanisms linking socioeconomic status and health status, no individual mechanism is so dominant that it alone is responsible for the bulk of the observed association. In other words, there may be different mechanisms underlying the association between socioeconomic status and health status over time. As Lutfey and Freese summarize, 'the association persists even while the relative influence of various proximate mechanisms changes' (Lutfey and Freese, 2005, p. 1328). It is in this sense that socioeconomic status is a 'fundamental cause' of health status, which is the key idea of the so-called theory of fundamental causes (Link and Phelan, 1995; Phelan et al., 2010).

It is evident that this case is difficult to characterize in terms of the pluralist approach. Indeed, Phelan, Link, and their associates do not take their study to establish two types of causation or two distinct causal claims. They contend that their study successfully identifies socioeconomic status as a 'fundamental cause' of health status (Link and Phelan, 1995, p. 80), which is not easily understood in a pluralist sense.

Suppose there are two different concepts of causality in the social sciences, say, difference-making causality ('causes$_{DM}$') and mechanistic causality ('causes$_{Mech}$'). If so, 'A causes$_{DM}$ B' says something different to 'A causes$_{Mech}$ B'. However, this is difficult to square with the study of socioeconomic status and health status. Link and Phelan (1995) have shown that there is a strong correlation between socioeconomic status and health status and there are some established mechanisms linking socioeconomic status to health status. What can we conclude from this? Is socioeconomic status a cause of health status? If so, in what sense?

1. Does the socioeconomic status cause$_{DM}$ health status?
2. Does socioeconomic status cause$_{Mech}$ health status?
3. Or, do Link and Phelan's studies suggest a new concept of causality?

It seems that the pluralist approach to causality in the social sciences leads to greater confusion. What is worse, the pluralist approach may lead to a problem of incommensurability. As Gerring argues,

If causation means different things to different people then, by definition, causal arguments cannot meet. If A says that X_1 caused Y and B retorts that it was, in fact, X_2 or that Y is not a proper outcome for causal investigation, and they claim to be basing their arguments on different understanding of causation, then these perspectives cannot be resolved; they are incommensurable.

(Gerring, 2005, p. 165)

Therefore, as Gerring (2005, p. 190) argues, 'pluralistic views are either unconvincing or, to the extent that they are true, unfortunate. We need a single framework within which to understand causal relationships in the social sciences.'

The epistemic theory provides the required simple and unified framework for understanding causal relationships in the social sciences. As we argued in §3.2, the epistemic theory obviates the need for a pluralist approach that invokes multiple concepts of cause. Even though epistemic causality invokes a single concept of cause, it can yield the correct causal judgements. This can be viewed as an instance of the argument from parsimony.

4.3 Causal Pluralism in the Social Sciences and the Argument from Evidential Pluralism

Phelan and Link's study fits epistemic causality well because it fits Evidential Pluralism: correlation and mechanism are established as a means to establish causation. This study is not an isolated case. In causal enquiry, it is not unusual that social scientists look for both types of evidence to support their causal claims, instead of focusing on one or the other. Other famous examples include Donohue and Levitt's study of legalized abortion and crimes, and Weinstein's study of rebellion and abortion (Shan and Williamson, 2021). As argued in §3.3, the fact that causal enquiry accords with Evidential Pluralism favours epistemic causality over rival accounts such as a difference-making monistic account or causal pluralism.

Phelan and Link's study can thus be used to exemplify both the argument from Evidential Pluralism and the argument from parsimony. Both the monistic approach and the pluralist approach have difficulties in accounting for the causal analysis of socioeconomic status and disease, while epistemic causality does not. Epistemic causality provides a more parsimonious understanding of Phelan and Link's study, and this study also accords well with Evidential Pluralism, which again favours epistemic causality.

Some might argue that in certain cases social scientists do indeed employ different methods to establish different causal claims. For example, in political science, some tend to use process tracing alone to make causal inferences, while others employ statistical techniques to establish causal claims. These methods are so different that one might reasonably infer that causal claims established using these methods appeal to different concepts of cause. If this is the case, the pluralist approach provides a better explanation than the epistemic theory (at least in some cases).

This argument is basically an inference from methodological diversity to causal pluralism (Maziarz, 2021).

Methodological diversity: there are different approaches to establishing causal claims in the social sciences.

Causal pluralism: there are different concepts of causality in the social sciences.

While the methodological diversity thesis seems to be true, causal pluralism is much more doubtful. Causal pluralism is certainly not required to explain methodological pluralism in the social sciences—epistemic causality provides an alternative account. The different methodological approaches to establishing causal claims can be understood as different ways to obtain evidence for monistic causal claims. Consider Weinstein's study of rebellion and violence. Weinstein (2007) argues for the causal claim that the initial conditions that rebel leaders encounter cause their strategy of violence. In order to support this causal claim, Weinstein integrates qualitative interview-based studies of the rebel groups and community-level social histories with statistical analysis of original newspaper datasets on patterns of violence in the case studies of rebel groups in Mozambique, Peru, and Uganda. In short, Weinstein uses both statistical techniques and ethnographic methods to establish a causal claim.

As argued by Shan and Williamson (2023), statistical techniques and ethnographic methods are not used to establish different types of causal claim. Rather they are used to obtain different objects of evidence to justify a causal claim. Therefore, methodological diversity does not support causal pluralism. Rather, it reflects Evidential Pluralism.

In addition, Shan and Williamson (2021) argue that even in the cases where political scientists are using process tracing alone to establish causal claims, evidence of correlation is assumed, though often implicitly. As Gerring (2005, p. 166) indicates, 'some correlational-style analyses slight the explicit discussion of causal mechanisms but this is usually because the author considers the causal mechanism to be clear and hence not worthy of explicit interrogation. Similarly, a mechanistic argument without any appeal to covariational patterns between X and Y does not make any sense. The existence of a causal mechanism presumes a pattern of association between a structural X and an ultimate Y.' Thus the use of process tracing methods does not imply that the concept of causality is mechanistic in nature. Nor does the use of statistical methods suggest that the concept of causality is fundamentally correlational. As we have argued, they can be understood as attempts to obtain different objects of evidence. As Crasnow (2011, p. 47) notes, 'pluralism about methodology need not commit us to a conceptual pluralism about causes'.

In sum, the epistemic theory provides an account of causality that fits well with causal enquiry in the social sciences. First, it does not fall to counterexamples that beset monistic accounts. Second, the epistemic theory provides a simpler and more unified account of causality than causal pluralism. Third, it fits very well with Evidential Pluralism, which provides a more tenable epistemological account of causality than causal pluralism.

5. Epistemic Causality in the Cognitive Sciences

Following Thagard (2005), we can define cognitive science as the interdisciplinary study of mind, embracing philosophy, psychology, artificial intelligence, neuroscience, linguistics, and anthropology. The explanatory results of cognitive science can then be understood as 'theoretical and experimental convergence on conclusions about the nature of mind', which—as with any science—must be framed in terms of empirically supported explanations and predictions.

A central task of cognitive science is to develop causal explanations (Cummins, 2000; Kaplan and Craver, 2011; Taylor, 2021). In some instances, cognitive theorists—such as psychologists—try to explain behaviour. For example, Piccinini and Craver (2011, 283) argue that:

> When psychologists explain behavior, the explanations typically make reference to causes that precede the behavior and make a difference to whether and how it occurs. For instance, they explain that Anna ducked because she saw a looming ball.

But cognitive theorists do not only try to explain behaviour; they also aim to develop causal explanations of the system—for example, the mind/brain—that is responsible for cognitive phenomena. In this vein, cognitive theorists have formulated a range of causal explanations of cognitive competences: for instance, causal explanations of categorization (cf. Davidoff, 2001; Harnad, 2017; Lin and Murphy, 1997; Taylor and Sutton, 2021), perception (Chater and Vitányi, 2003; Sims, 2018; Tanrıkulu et al., 2021), and memory (Baddeley, 1992; Michaelian and Sutton, 2013; Morrison and Chein, 2011).

Despite the focus on causal explanations in cognitive science, little attention has been paid to the epistemic theory of causality. Instead, the focus has been predominately on mechanistic (cf. Kaplan and Craver, 2011; Piccinini and Craver, 2011) and difference-making (e.g. interventionist) theories of causality (Meyer, 2020). We think that this is a mistake, because the epistemic theory of causality provides a very natural account of causality in cognitive science. Endorsing the epistemic theory of causality can help in at least two ways: first, to accommodate the epistemology of causality in cognitive science, which can be seen to conform to Evidential Pluralism (§5.1); second, to allow us to remain neutral about longstanding tensions in cognitive science concerning how to individuate causally efficacious mental states (§5.2).

5.1 Evidential Pluralism in Cognitive Science

We suggested in §2.3 that the epistemic theory of causality best accommodates Evidential Pluralism. Our claim here is that many working cognitive scientists conform to Evidential Pluralism and, as a result, the theory that best makes sense of their attempts to establish causal claims is the epistemic theory of causality.

As a reminder, Evidential Pluralism is the view that:

> In order to establish that A is a cause of B one normally needs to establish two things. First, that A and B are suitably correlated—typically, that A and B are probabilistically dependent, conditional on B's other known causes. Second, that there is some underlying mechanism linking A and B that can account for the difference that A makes to B.

Evidential Pluralism has already found good support in the health and social sciences, but open questions remain about its applicability elsewhere. Here, we argue that Evidential Pluralism is applicable to the cognitive sciences. To make this case, we consider two examples from cognitive neuroscience and developmental psychology respectively: Dehaene's (2009) theory of reading and discussions of 'theory of mind'.

Before turning to these examples, it is important to make one point explicit; namely, that in this subsection, we will be primarily concerned with demonstrating that Evidential Pluralism is, in fact, the causal epistemology that best describes cognitive scientific practice and, hence, that Evidential Pluralism is the causal epistemology that should be endorsed in (at least) this context. It follows from this that we have further support for the claim that the epistemic theory of causality provides the best account of causality in cognitive science, because we have already said (see §3.3 above) that the epistemic theory of causality accommodates Evidential Pluralism and that other theories of causality either are incompatible with Evidential Pluralism or have yet to provide an account of how they accommodate Evidential Pluralism.

Our first example is taken from Dehaene (2009, 5) who argues that 'the brain contains fixed circuitry exquisitely attuned to reading' and that the functional activity of this cortical area is causally responsible for our capacity to recognize words and letters. The area in question is located in the left ventral occipito-temporal junction and is now commonly labelled with a functional designation that

Dehaene himself coined: the visual word form area (VWFA). Dehaene's idea is that the function of VWFA is causally responsible for certain behaviours (namely, letter/word recognition and reading), because patterns of activity in VWFA play a causal role in the cognitive process that enables the organism to read by acting as a signal that informs the activities of downstream neural mechanisms.

Thus, Dehaene makes a causal claim: that patterns of activity in the VWFA (in response to certain environmental parameters) cause cognitive processes that enable organisms to recognize letters/words and, ultimately, to read. The open question, however, is how this causal claim is established. And it is here that Evidential Pluralism seems to be the causal epistemology at work. The reason is that Dehaene only feels able to put forward the aforementioned causal claim after establishing *both* mechanism and correlation.

In this case, the evidence of correlation is plentiful. For example, there is evidence that in normal literate subjects, VWFA is differentially responsive to written, but not spoken words (Dehaene and Cohen, 2007); that in blind subjects, the region is differentially responsive to words presented in Braille, but not to tactile control stimuli (Reich et al., 2011); and that lesions to VWFA appear to result in pure alexia, a condition in which formerly literate subjects cannot understand written words, despite being able to understand and produce verbal speech at roughly normal levels of competency (Gaillard et al., 2006). In each of these examples, we find that the functioning of VWFA (or not) makes a difference to our capacity to recognize letters/words and to read. This is exactly what Evidential Pluralism takes to be evidence of correlation.

But Dehaene also ensures that there is adequate evidence of mechanisms; specifically, evidence of the mechanisms for reading that link the functional activities of VWFA to the functional activities of other areas of the brain (see Figure 10.2). In particular, Dehaene (2009, 75) appeals to evidence of mechanisms from the mechanistic studies of Dehaene et al. (2002) in support of the following specific mechanism hypothesis:

> The left occipitotemporal "letterbox" [i.e. VWFA] identifies the visual form of letter strings. It then distributes this invariant visual information to numerous regions, spread over the left hemisphere, that encode word meaning, sound pattern, and articulation.... Learning to read thus consists in developing an efficient interconnection between visual areas and language areas. All connections are bidirectional.

This evidence of the organization and activities of different regions of the brain (e.g. posterior parietal region, occipital regions) linking VWFA to our capacity to read goes beyond mere association and has been supported by a number of mechanistic studies. In fact, in their conclusion of one such mechanistic study, Cohen et al. (2002, 1066) argue that:

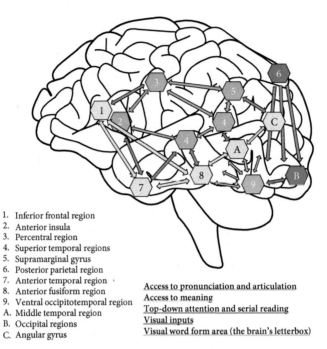

1. Inferior frontal region
2. Anterior insula
3. Percentral region
4. Superior temporal regions
5. Supramarginal gyrus
6. Posterior parietal region
7. Anterior temporal region
8. Anterior fusiform region
9. Ventral occipitotemporal region
A. Middle temporal region
B. Occipital regions
C. Angular gyrus

Access to pronunciation and articulation
Access to meaning
Top-down attention and serial reading
Visual inputs
Visual word form area (the brain's letterbox)

Figure 10.2 A modern vision of the cortical networks for reading (adapted from Dehaene, 2009, 84, fig. 2.2)

The fact that the location of the VWFA is highly reproducible across subjects suggests that some initial properties intrinsic to this region and to its pattern of connectivity are the cause of its subsequent specialization for reading.

We find, therefore, that Evidential Pluralism accords with the causal enquiries of working cognitive neuroscientists like Dehaene. But Evidential Pluralism also accords with the causal epistemology of another area of cognitive science: developmental psychology. In particular, the causal epistemology of Evidential Pluralism is at play in discussions about causal claims related to the 'theory of mind' (ToM) hypothesis.

According to the earliest account of ToM in Premack and Woodruff's (1978) study of the mind of chimpanzees:

> In saying that an individual has a theory of mind, we mean that the individual imputes mental states to himself and to others (either to conspecifics or to other species as well).   (Premack and Woodruff, 1978, 515)

ToM has been postulated to play a causal role in belief formation and action, which has been studied via so-called 'false-belief tasks', such as the Sally-Anne task described below:

Children are told a story in which Sally places a marble in a basket. Anne then moves the marble to a box while Sally is absent. Children are asked where Sally will look for her marble when she returns (action prediction) or simply where Sally thinks her marble is (belief). Normally developing children as young as 4 years typically pass such tasks, whereas children younger than 4 and much older children with autism typically fail. (Leslie et al., 2004, 515)

Thus, one (of many) causal claims involving ToM is that processes underlying ToM cause children to be successful at false-belief reasoning or that processes underlying ToM cause children to pass false-belief tasks such as the one described above.

To date, a large number of experiments involving false-belief tasks and their variants have been conducted and these seem to have shown that ToM emerges as causally efficacious mental states at around the age of 4. The problem, however, is that such experiments only provide evidence of a correlation between being a certain age (and, hence, possessing ToM) and success at false-belief tasks. This is the case because the results of these experiments are potentially subject to confounders that undermine the external and internal validity of the results.

For example, some have argued that we must be cautious when generalizing the results of experiments involving false-belief tasks to other domains, because such experiments have focused almost exclusively on subjects from WEIRD (Western, Educated, Industrialized, Rich, and Democratic) societies (Heinrich et al., 2010). Moreover, some have argued that the results of experiments involving false-belief tasks do not adequately test for the causal role of ToM, because standard false-belief tasks are 'verbally based' and so we cannot rule out the possibility that the results are biased by factors related to children's linguistic abilities (Southgate et al., 2007; Onishi and Baillargeon, 2005; Surian et al., 2007). The point, then, is that the experimental results of false-belief only provide strong evidence along channel α_1 of Figure 10.3 (where A is being a certain age (and, hence, possessing ToM) and B is success at false-belief tasks).

Now, if correlation were sufficient for causality, the evidence of correlation obtained via experiments involving false-belief tasks should be enough to establish the causal claim that being a certain age (and, hence, possessing ToM) (A) causes success at false-belief tasks (B). But this is certainly not how practising developmental psychologists have seen the matter. Instead, they have been reluctant to assert that the causal claim has been established until they have acquired evidence of mechanisms. This is clear if we take a brief look at the recent history of research related to theory of mind mechanisms (ToMM). According to Frith and Happé (1999, 82):

The cognitive processes which underlie the development of Theory of Mind (ToM) are still a matter of debate. The field can be divided into those who favour

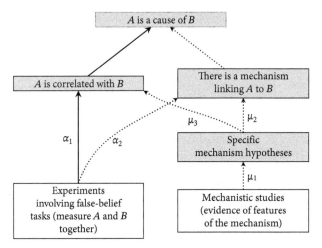

Figure 10.3 Evidence from experiments of a correlation between being a certain age (and, hence, possessing ToM) and success at false-belief tasks

a more general explanation for ToM (e.g. simulation, general theory building), and those who argue for the necessity of a dedicated cognitive mechanism (for debate see, for example, Goldman, 1993; Gopnik, 1993; and chapters in Carruthers and Smith, 1996).

Both camps accept the results of experiments involving false-belief tasks, but they disagree about the nature of the cognitive mechanisms underlying ToM that link it to success at false-belief tasks. For some, the mechanisms underlying ToM are 'domain general' in the sense that they are not specifically and uniquely tied to our ToM-based ability to attribute mental states to others and predict their behaviour accordingly. For others, however, there is a dedicated—and, perhaps, innately specified—cognitive mechanism underlying ToM.

As an example, consider the claim by Leslie (1987) that 'basic representational structures for a theory of mind are put in place by the emergence of the decoupler mechanism'. This view—which falls into the domain-specific camp—holds that the origins of ToM can be found in various 'decoupling mechanisms', which underpin the ability to represent mental states 'decoupled' from reality. An example is the decoupling mechanism for pretence (see Figure 10.4), which, according to Leslie (1987, 419–420), operates as follows:

First, there are the perceptual processes whose job is to feed representations of the current situation to the central processes. Second, there is the set of processes labeled central cognitive systems. These include structures corresponding to perceived situation, memory systems (including, for example, general knowledge), systems for planning action, and so on....

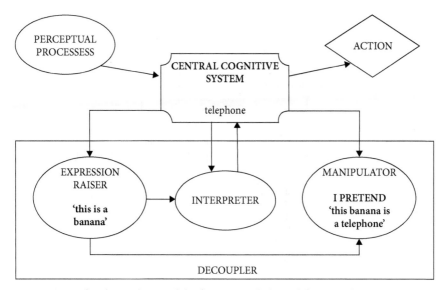

Figure 10.4 The decoupler model of pretence (adapted from Leslie, 1987, 419, fig. 2)

The expression raiser's job is to copy primary representations from the central systems. It raises copies into the opaque context of the decoupling marks. The copy of the primary expression is thus removed from its normal input-output relations and from the normal computational consequences it would otherwise have. It will now form the nucleus of a metarepresentation...

The manipulator's job is to transform decoupled expressions by integrating (primary) information from memory within the decoupling marks or by applying inference rules from memory....

The interpreter can access primary representations in central systems. It performs anchoring functions and relates decoupled expressions to the current perceptual representation. It can access inference rules and other information for passing to the manipulator in a further cycle.

For Leslie, it is the emergence of metarepresentation through the decoupling mechanism that accounts for abilities associated with ToMM and, ultimately, for success at false-belief tasks.

Leslie's account counts as a mechanism hypothesis in our sense and this hypothesis has now found support from several functional neuroimaging studies of ToMM, which purport to identify the neural substrates of 'mentalizing' abilities associated with ToM. According to a review by Gallagher and Frith (2003, 78), 'the findings of these studies indicate that this ability is mediated by a highly circumscribed region of the brain, the anterior paracingulate cortex (approximately

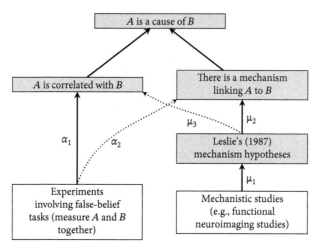

Figure 10.5 Establishing that being a certain age (and, hence, possessing ToM) causes success at false-belief tasks

corresponding to Brodmann area (BA) 9/32)'. Moreover, evidence has now been gathered to link Leslie's ToMM to other relevant mechanisms of inhibitory selection, where certain default beliefs are inhibited so that other beliefs can be selected.

Thus, those in favour of the view that there are dedicated and, perhaps, domain-specific cognitive mechanisms underlying ToM will likely assume that we have evidence of a mechanism linking being a certain age (and, hence, possessing ToM) (A) and success at false-belief tasks (B) (see Figure 10.5). And this counts as evidence of mechanisms along channels μ_1 and μ_2, which combines with evidence of correlation along channel α_1 and so allows us to establish that A is a cause of B. This is exactly why Leslie (1987, 423) asserts that her account 'builds a powerful causal story' by building upon the evidence garnered from experiments involving false-belief tasks.

That said, open questions remain about whether it is right to think of ToMM in the dedicated and domain-specific way Leslie favours. For those who are sceptical of this approach, it seems unlikely that the evidence of mechanisms cited above would be seen as definitive. Instead, advocates of domain-general views of ToMM typically call for further research to move beyond the state of affairs represented by Figure 10.3 above. For example, research to study 'tasks based on a refined analysis of specific component processes of mentalizing..., rather than fixating on the umbrella term ToM' (Schurz and Perner, 2015, 1610).

This demonstrates that for advocates of both domain-general and domain-specific ToMM, the evidence of mechanisms can never be enough to establish causal claims on its own. Every theorist in this area refers back to experimental results delivering evidence of correlation, because without referring to these results there is nothing to determine whether there is a net correlation. This follows because we cannot undertake, for instance, functional neuroimaging without first specifying the task that participants are to be doing while we scan their brains. And the only way to identify the relevant tasks in this domain is to first obtain evidence of a correlation between some putative cause A—for example, being a certain age (and, hence, possessing ToM)—and some effect B—for example, success at false-belief tasks.

Thus, the lesson is clear: developmental psychologists studying ToM are not willing to assert that a causal claim has been established until they have evidence of *both* correlation and mechanism. Neither will be sufficient in isolation. This accords perfectly with Evidential Pluralism. Moreover, Evidential Pluralism favours the epistemic theory of causality over rival theories. Neither the difference-making nor mechanistic theories of causality seem to make sense of this kind of causal enquiry in developmental psychology, because under a difference-making theory the search for mechanisms is inexplicable, and under a mechanistic theory the original focus on gaining evidence of correlation is inexplicable.

5.2 Neutrality in Cognitive Sciences

In line with the argument from neutrality presented in §2.4, we see next that by endorsing the epistemic theory of causality we can remain neutral about long-standing and intractable tensions in cognitive science. More precisely, we argue that by endorsing the epistemic theory we can, in some cases at least, remain neutral about how to individuate causally efficacious mental states.

To understand how endorsing the epistemic theory of causality can allow us to remain neutral about how to individuate causally efficacious mental states, it is helpful to consider an example: Marr's (1982) computational theory of vision. According to Marr's theory, perceptual mechanisms solve information-processing tasks set to them by nature with the aim of deriving a representation of three-dimensional shape from information contained in two-dimensional images. Egan (1992, 453) gives a helpful recapitulation of this mechanism as follows:

> Marr's theory divides this task into three distinct stages, each involving the construction of a representation, tokens of which serve as inputs to subsequent processes. Vision culminates in a representation that is suitable for the recognition of objects. Innate assumptions...incorporated into the visual system itself, and reflecting physical constraints on the pairing of retinal images with distal shapes, allow the postulated mechanisms underlying early vision to recover information about the distal scene based only on information contained in the image.

To explain how the visual system undertakes this task, Marr's theory specifies a range of functions that are computed by the visual system—for example, the following function that characterizes how the visual system initially filters the image:

$$\nabla^2 G * I(x, y) \tag{1}$$

Following Marr and Hildreth (1980), $\nabla^2 G$ is taken to represent a filter that 'detect[s] intensity changes efficiently' in virtue, for example, of being:

> capable of being tuned to act at any desired scale, so that large filters can be used to detect blurry shadow edges, and small ones to detect sharply focused fine detail in the image.   (Marr, 1982, 54)[10]

$I(x,y)$ represents the image to be filtered, and $*$ the operation of convolution (which serves to determine the most important portions of an image) (Marr, 1982, 54–8, 338). The details here are complex, but all that matters is that, for Marr, this formal characterization is, from a 'computational point of view', a 'precise speci-fication of what the retina does' when initially filtering an image (see Figure 10.6).

Now, most cognitive theorists accept that Marr's computational explanation of vision is a causal explanation, but there has been heated debate about the locus of causality in this instance. Some argue that causal relations obtain between the contents of the representations over which the system computes: for example, that the system produces an early representation R_1 and the content of R_1 is causally efficacious in the production of later representations $R_2, \ldots, R_n$. Others, however, argue that causal relations obtain between physical properties of the system that realizes the functional (read: representational) states. Egan (1992, 446) takes this view when she argues that representations (as symbols):

> are just functionally characterized objects whose individuation conditions are specified by a realization function f_R which maps equivalence classes of physical features of a system to what we might call "symbolic" features. Formal operations are just those physical operations that are differentially sensitive to the aspects of symbolic expressions that under the realization function f_R are specified as symbolic features. The mapping f_R allows a causal sequence of physical state transitions to be interpreted as a *computation*.

---

[10] According to Marr (1982, 55), $\nabla^2 G$ is a 'circularly symmetric Mexican hat-shaped operator whose distribution in two dimensions may be expressed in terms of the radial distance r from the origin by the formula:

$$\nabla^2 G(r) = -\frac{1}{\pi \sigma^4} \left(1 - \frac{r^2}{2\sigma^2} \right) \exp\left(\frac{r^2}{\sigma^2} \right)'$$

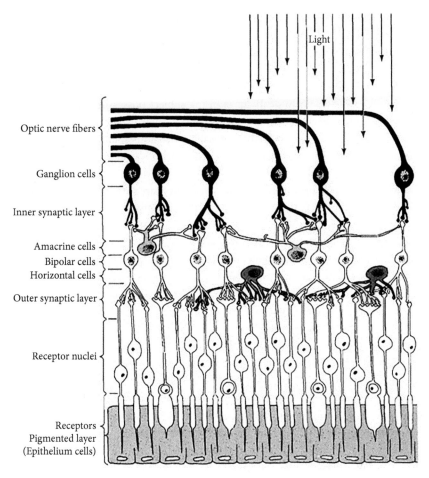

Figure 10.6 A cross-section of the retina, part of whose function is to compute (1) (from Marr, David, foreword by Shimon Ullman, afterword by Tomaso Poggio, *Vision*, Fig. 7.1 (p. 338), 2010, Lucia M. Vaina, by permission of The MIT Press)

At stake here is the question of whether or not the various modules of the visual system (i.e. what the system does) are 'individuated essentially by reference to the contents of the representational tokens that form the inputs and outputs of these modules' (Egan, 1992, 453). Egan takes the view that they are not, because the realization function f_R should be understood as individuating computational states *non-semantically*. The standard view, however, is that such modules are individuated by reference to the contents of representational tokens, because intentional (read: semantic) mental states are individuated by their contents. Thus, there is tension between semantic and non-semantic individuations of the causally efficacious mental states involved in vision.

But this is not the only tension concerning the individuation of the causally efficacious mental states involved in vision, because even those who defend a semantic individuation of such states do not agree about whether the contents that individuate these states entirely supervene on intrinsic (physical) states of the subject possessing them. Some argue that the individuating contents do entirely supervene on intrinsic (physical) states of the subject possessing them (so-called internalists), but others argue that they do not (so-called externalists). This second tension, therefore, is about whether causally efficacious mental states are individuated (in part) by reference to the social and physical environment of the subject possessing them.

What is important here is that all can agree that Marr's computational theory of vision provides a good causal explanation of the cognitive competency in question (e.g. vision), while disagreeing about which mental states are causally efficacious (non-semantic physical states vs representational states) or about the supervenience base of causally efficacious, representational mental states (intrinsic (physical) states vs (partly) social and physical environment). As a consequence, there is uncertainty about what exactly Marr's theory is a good causal explanation of: non-semantic physical states, representational states with contents supervening only on the intrinsic (physical) states of the subjects possessing them, or representational states with contents supervening (in part) on the social or physical environment.

Importantly, however, Marr argues that a good explanation of vision will incorporate descriptions at three 'logically and causally related' levels: the level of 'computational theory' (specifying the computed function), the level of representation and algorithm (describing how the function is computed), and the level of hardware implementation (describing the neural states supporting the computation). As such, Marr's theory is neutral about how to individuate the causally efficacious mental states of vision, because he asserts that a complete causal explanation of vision will refer both to 'neural mechanisms' at the level of hardware implementation and, at the level of computational theory, to 'channels' in the 'visual pathways' that detect spatial patterns 'based on a form of spatial probability summation' (Marr, 1982, 62).[11]

And this is where the problem lies, because insofar as we endorse a difference-making or mechanistic theory of causality, we cannot remain neutral about how to individuate the causally efficacious mental states. To see why, note first that all of the possible interpretations of the relevant causally efficacious mental states can be said to stand in difference-making or mechanistic causal relations. For example, it is just as possible that there is an appropriate sort of mechanism linking non-semantic physical states as it is that there is an appropriate sort of mechanism

---

[11] In fact, it is because Marr's theory is neutral in this way that philosophers have been able to disagree about how the causally efficacious mental states are to be individuated (cf. Burge, 1986; Egan, 1991; Kitcher, 1988; Segal, 1991).

linking representational states, and this holds true however you think the contents of representational states are individuated. Likewise, it is just as possible that, say, a chain of counterfactual dependence runs between non-semantic physical states as it is that a chain of counterfactual dependence runs between representational states.

The problem, however, is that mechanistic and difference-making theories of causality cannot remain neutral in this regard without incurring the charge of being vacuous. This is the case because it is only possible to spell out the mechanistic or difference-making details when we take a view on how to individuate the causal relata, since the mechanisms or relations of difference-making will be different if we take these relata to be non-semantic physical states or representational states respectively. The key point, therefore, is that it is not possible for mechanistic and difference-making theories of causality to say simply that a computational state, A, caused some effect, B (e.g. the filtering of an image), without specifying how A is individuated, because then the relevant mechanism or difference-making relations are themselves left unspecified.

The epistemic theory of causality does not suffer from the same problems, because we need not take a view on how to individuate the relevant causally efficacious mental states in order to have a causal belief of the form $A \rightarrow B$. In this way, the epistemic theory need not prejudge what is, ultimately, an open question in the sciences of the mind. This follows because the epistemic theory of causality can cope with any possible individuation and, unlike mechanistic and difference-making theories of causality, need not take a stance on this issue while we remain unsure about which individuation is correct.[12] So, on the epistemic view, Marr's theory can be said to support a set of causal beliefs about causally efficacious computational states that allow us to predict, explain, and control a particular portion of reality; namely, the cognitive system(s) responsible for vision. As such, we need only argue that Marr's theory supports causal beliefs of the form $A \rightarrow B$, where A is, say, the transitions of those computational states (however they are individuated) carrying out function (1) above and B is the visual system's filtering of the image.

It is clear in practice that Marr's theory does support a range of causal beliefs of the form $A \rightarrow B$ that motivate successful prediction, explanation, and control inferences. For example, it supports the causal belief that our perceptual mechanisms (A_1) cause images processed on two separate retina to be 'fused' if they occupy a region in visual space known as 'Panum's fusional area' (B_1), which motivates the prediction that if images processed do not occupy Panum's fusional area, then we will lose the perception of these objects as being a single unified object and will see

---

[12] Of course, it might still matter in the end how the relevant causally efficacious mental states are individuated, because the causal belief $A \rightarrow B$ might lead to successful PECs under one individuation but not under another.

instead two images of the same object. Furthermore, it supports the causal belief that disjunctive eye movement (A_2) causes changes in the plane of fixation (B_2), which motivates the control inference that if we prevent the eyes from moving disjunctively, then we prevent the visual system from changing the plane of fixation (by preventing the two lines of sight from converging or diverging).

It is clear that many working cognitive scientists do take Marr's theory to offer a causal explanation of vision, but this is obfuscated by the philosophical debate about how to individuate causally efficacious mental states. If we endorse the mechanistic or difference-making theories of causality, we cannot leave this debate behind. However, if we endorse the epistemic theory of causality, then we can agree that Marr's computational theory of vision provides a good causal explanation without getting bogged down in such debates. Thus, the epistemic theory will be appealing to those who are interested in defending a theory of causality that is able to prioritize scientific practice and consensus by remaining neutral about philosophical issues such as individuation.

6. Conclusion

Epistemic causality is a theory of the nature of causality, but one that gives primacy to the epistemology of causality. Causality is analysed in terms of rational causal beliefs, and the question of which causal beliefs are rational is one of causal epistemology. This tight connection between metaphysics, conceptual analysis, and epistemology makes the epistemic theory an account of causality that is particularly close to scientific practice. We have seen that a careful consideration of both the social and cognitive sciences lends credibility to epistemic causality. Our examples can be thought of as instantiating four arguments for epistemic causality: arguments from failure, parsimony, Evidential Pluralism, and neutrality. We appealed to the social sciences to illustrate the arguments from failure, parsimony, and Evidential Pluralism, and we appealed to the cognitive sciences to illustrate the arguments from Evidential Pluralism and neutrality.

Taking these arguments together with those of Russo and Williamson (2007) for epistemic causality in the health sciences, one can make the case that epistemic causality coheres well with causal practices across a broad range of sciences. This in itself is an important virtue of a theory of causality.

Acknowledgements

Some material from §4 appeared previously in Shan and Williamson (2023).

This research was supported by funding from the Leverhulme Trust (grants RPG-2019-059 and RPG-2022-336) and the British Academy (grant SRG1920/101076).

References

Adler, N. E. and Newman, K. (2002). Socioeconomic disparities in health: Pathways and policies. *Health Affairs*, 21: 60–76.

Adler, N. E., Boyce, T., Chesney, M. A., Cohen, S., Folkman, S., Kahn, R. L., and Syme, S. L. (1994). Socioeconomic status and health: The challenge of the gradient. *American Psychologist*, 49(1): 15–24.

Ahn, W., Kalish, C. W., Medin, D. L., and Gelman, S. A. (1995). The role of covariation versus mechanism information in causal attribution. *Cognition*, 54(3): 299–352.

Baddeley, A. (1992). Working memory: The interface between memory and cognition. *Journal of Cognitive Neuroscience*, 4(3): 281–8.

Beach, D. and Pedersen, R. B. (2013). *Process-Tracing Methods*. University of Michigan Press, Ann Arbor, MI.

Beebee, H. (2015). Causation, projection, inference and agency. In Johnson, R. N. and Smith, M., editors, *Passions and Projections: Themes from the Philosophy of Simon Blackburn*, pages 25–48. Oxford University Press, Oxford.

Bosworth, B. (2018). Increasing disparities in mortality by socioeconomic status. *Annual Review of Public Health*, 39(1): 237–51.

Burge, T. (1986). Individualism and psychology. *Philosophical Review*, 95(1): 3–45.

Carruthers, P. and Smith, P. K. (1996). *Theories of Theories of Mind*. Cambridge University Press, Cambridge.

Chater, N. and Vitányi, P. (2003). Simplicity: A unifying principle in cognitive science? *Trends in Cognitive Sciences*, 7(1): 19–22.

Cohen, L., Lehéricy, S., Chochon, F., Lemer, C., Rivaud, S., and Dehaene, S. (2002). Language-specific tuning of visual cortex? Functional properties of the visual word form area. *Brain*, 125(5): 1054–69.

Cook, T. D. and Campbell, D. T. (1979). *Quasi-experimentation: Design and Analysis Issues for Field Settings*. Rand McNally, Chicago, IL.

Crasnow, S. (2011). Evidence for use: Causal pluralism and the role of case studies in political science research. *Philosophy of the Social Sciences*, 41(1): 26–49.

Cummins, R. C. (2000). 'How does it work' versus 'what are the laws?': Two conceptions of psychological explanation. In Keil, F. and Wilson, R. A., editors, *Explanation and Cognition*, pages 117–45. MIT Press, Cambridge, MA.

Davidoff, J. (2001). Language and perceptual categorisation. *Trends in Cognitive Sciences*, 5(9): 382–7.

Dehaene, S. (2009). *Reading in the Brain: The New Science of How We Read*. Penguin, New York.

Dehaene, S. and Cohen, L. (2007). Cultural recycling of cortical maps. *Neuron*, 56(2): 384–98.

Dehaene, S., Le Clec'H, G., Poline, J., Le Bihan, D., and Cohen, L. (2002). The visual word form area: A prelexical representation of visual words in the fusiform gyrus. *Neuroreport*, 13(3): 321–5.

Donohue, J. J. and Levitt, S. D. (2001). The impact of legalized abortion on crime. *Quarterly Journal of Economics*, 116(2): 379–420.

Dutton, D. B. (1986). Social class, health, and illness. In Aiken, L. and Mechanic, D., editors, *Applications of Social Science to Clinical Medicine and Health Policy*, pages 31–62. Rutgers University Press, New Brunswick, NJ.

Egan, F. (1991). Must psychology be individualistic? *Philosophical Review*, 100(2): 179–203.

Egan, F. (1992). Individualism, computation, and perceptual content. *Mind*, 101(403): 443–59.

Evans, G. W. and Saegert, S. (2000). Residential crowding in the context of inner city poverty. In Wapner, S., Demick, J., Yamamoto, T., and Minami, H., editors, *Theoretical Perspectives in Environment-Behavior Research: Underlying Assumptions, Research Problems, and Methodologies*, pages 247–67. Springer US, Boston, MA.

Frith, U. and Happé, F. (1999). Theory of mind and self-consciousness: What is it like to be autistic? *Mind and Language*, 14(1): 82–9.

Gaillard, R., Naccache, L., Pinel, P., Clémenceau, S., Volle, E., Hasboun, D., Dupont, S., Baulac, M., Dehaene, S., Adam, C., et al. (2006). Direct intracranial, FMRI, and lesion evidence for the causal role of left inferotemporal cortex in reading. *Neuron*, 50(2): 191–204.

Gallagher, H. L. and Frith, C. D. (2003). Functional imaging of 'theory of mind'. *Trends in Cognitive Sciences*, 7(2): 77–83.

Gerring, J. (2005). Causation: A unified framework for the social sciences. *Journal of Theoretical Politics*, 17(2): 163–98.

Gerstenberg, T., Goodman, N. D., Lagnado, D. A., and Tenenbaum, J. B. (2022). A counterfactual simulation model of causal judgments for physical events. *Psychological Review*, 128(5): 936–75.

Glynn, A. N. and Ichino, N. (2015). Using qualitative information to improve causal inference. *American Journal of Political Science*, 59(4): 1055–71.

Goertz, G. and Mahoney, J. (2012). *A Tale of Two Cultures*. Princeton University Press, Princeton and Oxford.

Goldman, A. I. (1993). The psychology of folk psychology. *Behavioral and Brain Sciences*, 16(1): 15–28.

Goldthorpe, J. H. (2001). Causation, statistics, and sociology. *European Sociological Review*, 17(1): 1–20.

Gopnik, A. (1993). How we know our minds: The illusion of first-person knowledge of intentionality. *Behavioral and Brain Sciences*, 16(1): 1–14.

Gopnik, A., Glymour, C., Sobel, D. M., Schulz, L. E., Kushnir, T., and Danks, D. (2004). A theory of causal learning in children: Causal maps and Bayes nets. *Psychological Review*, 111(1): 3–32.

Gould, E. D., Weinberg, B. A., and Mustard, D. B. (2002). Crime rates and local labor market opportunities in the United States: 1979–1997. *Review of Economics and Statistics*, 84(1): 45–61.

Granger, C. W. J. (1969). Investigating causal relations by econometric models and cross-spectral methods. *Econometrica*, 37(3): 424–38.

Granger, C. W. J. (1980). Testing for causality. *Journal of Economic Dynamics*, 2: 329–52.

Guyatt, G., Cairns, J., Churchill, D., Cook, D., Haynes, B., Hirsh, J., Irvine, J., Levine, M., Levine, M., Nishikawa, J., Sackett, D., Brill-Edwards, P., Gerstein, H., Gibson, J., Jaeschke, R., Kerigan, A., Neville, A., Panju, A., Detsky, A., Enkin, M., Frid, P., Gerrity, M., Laupacis, A., Lawrence, V., Menard, J., Moyer, V., Mulrow, C., Links, P., Oxman, A., Sinclair, J., and Tugwell, P. (1992). Evidence-based medicine: A new approach to teaching the practice of medicine. *Journal of the American Medical Association*, 268(17): 2420–5.

Haggard, S. and Kaufman, R. R. (2016). *Dictators and Democrats*. Princeton University Press, Princeton and Oxford.

Hall, N. (2004). Two concepts of causation. In Collins, J., Hall, N., and Paul, L., editors, *Causation and Counterfactuals*, pages 225–76. MIT Press, Cambridge, MA and London.

Harnad, S. (2017). To cognize is to categorize: Cognition is categorization. In Cohen, H. and Lefebvre, C., editors, *Handbook of Categorization in Cognitive Science*, pages 21–54. Elsevier, New York.

Hedström, P. and Ylikoski, P. (2010). Causal mechanisms in the social sciences. *Annual Review of Sociology*, 36(1): 49–67.

Heinrich, J., Heine, S. J., and Norenzayan, A. (2010). The weirdest people in the world. *Behavioral and Brain Sciences*, 33(2–3): 61–83.

Holland, P. W. (1986). Statistics and causal inference. *Journal of the American Statistical Association*, 81(396): 945–60.

House, J. S., Lepkowski, J. M., Kinney, A. M., Mero, R. P., Kessler, R. C., and Herzog, A. R. (1994). The social stratification of aging and health. *Journal of Health and Social Behavior*, 35(3): 213–34.

Hume, D. (1748). *An Enquiry Concerning Human Understanding*. Oxford Philosophical Texts. Oxford University Press, Oxford.

Illsley, R. and Mullen, K. (1985). The health needs of disadvantaged client groups. In Holland, W. W., Detels, R., and Knox, G., editors, *Oxford Textbook of Public Health*, pages 389–402. Oxford University Press, Oxford.

Imai, K. (2017). *Quantitative Social Science: An Introduction*. Princeton University Press, Princeton.

Jaynes, E. T. (2003). *Probability Theory: The Logic of Science*. Cambridge University Press, Cambridge.

Kaplan, D. and Craver, C. F. (2011). The explanatory force of dynamical and mathematical models in neuroscience: A mechanistic perspective. *Philosophy of Science*, 78(4): 601–27.

King, G., Keohane, R. O., and Verba, S. (1994). *Designing Social Inquiry*. Princeton University Press, Princeton.

Kitcher, P. (1988). Marr's computational theory of vision. *Philosophy of Science*, 55(1): 1–24.

Lagnado, D. A. and Sloman, S. (2004). The advantage of timely intervention. *Journal of Experimental Psychology: Learning, Memory and Cognition*, 30(4), 856–76.

Lantz, P. M., House, J. S., Lepkowski, J. M., Williams, D. R., Mero, R. P., and Chen, J. (1998). Socioeconomic factors, health behaviors, and mortality: Results from a nationally representative prospective study of US adults. *Journal of the American Medical Association*, 279(21): 1703–8.

Leslie, A. M. (1987). Pretense and representation: The origins of 'theory of mind'. *Psychological Review*, 94(4): 412.

Leslie, A. M., Friedman, O., and German, T. P. (2004). Core mechanisms in 'theory of mind'. *Trends in Cognitive Sciences*, 8(12): 528–33.

Levitt, S. D. (2004). Understanding why crime fell in the 1990s: Four factors that explain the decline and six that do not. *Journal of Economic Perspectives*, 18(1): 163–90.

Lewis, D. (1973). Causation. *Journal of Philosophy*, 70: 556–67.

Lin, E. L. and Murphy, G. L. (1997). Effects of background knowledge on object categorization and part detection. *Journal of Experimental Psychology: Human Perception and Performance*, 23(4): 1153–69.

Link, B. G. and Phelan, J. C. (1995). Social conditions as fundamental causes of disease. *Journal of Health and Social Behavior* (Extra issue): 80–94.

Longworth, F. (2006). Causation, pluralism and responsibility. *Philosophica*, 77: 45–68.

Lutfey, K. and Freese, J. (2005). Toward some fundamentals of fundamental causality: Socioeconomic status and health in the routine clinic visit for diabetes. *American Journal of Sociology*, 110(5): 1326–72.

Machin, S. and Meghir, C. (2004). Crime and economic incentives. *Journal of Human Resources*, 39(4): 958–79.

Marr, D. (1982). *Vision: A Computational Investigation into the Human Representation and Processing of Visual Information*. W. H. Freeman, New York.

Marr, D. and Hildreth, E. (1980). Theory of edge detection. *Proceedings of the Royal Society of London*, Series B, 207: 187–217.

Maziarz, M. (2020). *The Philosophy of Causality in Economics: Causal Inferences and Policy Proposals*. Routledge, London and New York.

Maziarz, M. (2021). Resolving empirical controversies with mechanistic evidence. *Synthese*, 199: 9957–78.

Meyer, R. (2020). The non-mechanistic option: Defending dynamical explanations. *British Journal for the Philosophy of Science*, 71(3): 959–85.

Michaelian, K. and Sutton, J. (2013). Distributed cognition and memory research: History and current directions. *Review of Philosophy and Psychology*, 4(1): 1–24.

Morgan, S. L. and Winship, C. (2015). *Counterfactuals and Causal Inferences: Methods and Principles for Social Research*, 2nd ed. Cambridge University Press, Cambridge.

Morrison, A. B. and Chein, J. M. (2011). Does working memory training work? The promise and challenges of enhancing cognition by training working memory. *Psychonomic Bulletin and Review*, 18(1): 46–60.

Onishi, K. H. and Baillargeon, R. (2005). Do 15-month-old infants understand false beliefs? *Science*, 308(5719): 255–8.

Pampel, F. C., Krueger, P. M., and Denney, J. T. (2010). Socioeconomic disparities in health behaviors. *Annual Review of Sociology*, 36: 349–70.

Parkkinen, V.-P., Wallmann, C., Wilde, M., Clarke, B., Illari, P., Kelly, M. P., Norell, C., Russo, F., Shaw, B., and Williamson, J. (2018). *Evaluating Evidence of Mechanisms in Medicine: Principles and Procedures*. Springer, Cham.

Patton, M. Q. (2002). *Qualitative Research and Evaluation Methods*, 3rd ed. Sage, Thousand Oaks, CA.

Phelan, J. C., Link, B. G., Diez-Roux, A., Kawachi, I., and Levin, B. (2004). 'Fundamental causes' of social inequalities in mortality: A test of the theory. *Journal of Health and Social Behavior*, 45: 265–85.

Phelan, J. C., Link, B. G., and Tehranifar, P. (2010). Social conditions as fundamental causes of health inequalities: Theory, evidence, and policy implications. *Journal of Health and Social Behavior*, 51 Suppl: S28–S40.

Piccinini, G. and Craver, C. (2011). Integrating psychology and neuroscience: Functional analyses as mechanism sketches. *Synthese*, 183(3): 283–311.

Premack, D. and Woodruff, G. (1978). Does the chimpanzee have a theory of mind? *Behavioral and Brain Sciences*, 1(4): 515–26.

Reich, L., Szwed, M., Cohen, L., and Amedi, A. (2011). A ventral visual stream reading center independent of visual experience. *Current Biology*, 21(5): 363–8.

Reiss, J. (2009). Causation in the social sciences: Evidence, inference, and purposes. *Philosophy of the Social Sciences*, 39(1): 20–40.

Reiss, J. (2012). Causation in the sciences: An inferentialist account. *Studies in History and Philosophy of Science Part C: Studies in History and Philosophy of Biological and Biomedical Sciences*, 43(4): 769–77.

Rohlfing, I. and Zuber, C. I. (2021). Check your truth conditions! Clarifying the relationship between theories of causation and social science methods for causal inference. *Sociological Methods and Research*, 50(4): 1623–59.

Rothman, K. (1986). *Modern Epidemiology*. Little, Brown, and Company, Boston, MA.

Rubin, D. B. (1974). Estimating causal effects of treatments in randomized and nonrandomized studies. *Journal of Educational Psychology*, 66: 688–701.

Russo, F. (2009). *Causality and Causal Modelling in the Social Sciences: Measuring Variations*. Springer, New York.

Russo, F. and Williamson, J. (2007). Interpreting causality in the health sciences. *International Studies in the Philosophy of Science*, 21(2): 157–70.

Russo, F. and Williamson, J. (2011a). Epistemic causality and evidence-based medicine. *History and Philosophy of the Life Sciences*, 33(4): 563–82.

Russo, F. and Williamson, J. (2011b). Generic versus single-case causality: The case of autopsy. *European Journal for Philosophy of Science*, 1(1): 47–69.

Sackett, D. L., Rosenberg, W. M. C., Gray, J. A. M., Haynes, R. B., and Richardson, W. S. (1996). Evidence based medicine: What it is and what it isn't. *British Medical Journal*, 312(7023): 71–2.

Schnall, P. L., Pieper, C., Schwartz, J. E., Karasek, R. A., Schlussel, Y., Devereux, R. B., Ganau, A., Alderman, M., Warren, K., and Pickering, T. G. (1990). The relationship between 'job strain,' workplace diastolic blood pressure, and left ventricular mass index: Results of a case-control study. *Journal of the American Medical Association*, 263(14): 1929–35.

Schurz, M. and Perner, J. (2015). An evaluation of neurocognitive models of theory of mind. *Frontiers in Psychology*, 6: 1610.

Segal, G. (1991). Defence of a reasonable individualism. *Mind*, 100(4): 485–94.

Shan, Y. and Williamson, J. (2021). Applying evidential pluralism to the social sciences. *European Journal for the Philosophy of Science*, 11(4): 96.

Shan, Y. and Williamson, J. (2023). *Evidential Pluralism in the Social Sciences*. Routledge, London and New York.

Sims, C. R. (2018). Efficient coding explains the universal law of generalization in human perception. *Science*, 360(6389): 652–6.

Sobel, D. M. and Kushnir, T. (2006). The importance of decision making in causal learning from interventions. *Memory and Cognition*, 34(2): 411–19.

Southgate, V., Senju, A., and Csibra, G. (2007). Action anticipation through attribution of false belief by 2-year-olds. *Psychological Science*, 18(7): 587–92.

Surian, L., Caldi, S., and Sperber, D. (2007). Attribution of beliefs by 13-month-old infants. *Psychological Science*, 18(7): 580–6.

Tanrıkulu, Ö. D., Chetverikov, A., Hansmann-Roth, S., and Kristjánsson, Á. (2021). What kind of empirical evidence is needed for probabilistic mental representations? An example from visual perception. *Cognition*, 217: 104903.

Taylor, S. D. (2021). Causation and cognition: An epistemic approach. *Synthese*, 199: 9133–60.

Taylor, S. D. and Sutton, P. R. (2021). A frame-theoretic model of Bayesian category learning. In *Concepts, Frames and Cascades in Semantics, Cognition and Ontology*, Löbner, S., Gamerschlag, T., Kalenscher, T., Schrenk, M., and Zeevat, H., editors, pages 329–49. Springer, Cham.

Teddlie, C. and Tashakkori, A. (2009). *Foundations of Mixed Methods Research: Integrating Quantitative and Qualitative Approaches in the Social and Behavioral Sciences*. Sage, Thousand Oaks, CA.

Thagard, P. (2005). *Mind: Introduction to Cognitive Science*. MIT Press, Cambridge, MA.

Weber, E. (2009). How probabilistic causation can account for the use of mechanistic evidence. *International Studies in the Philosophy of Science*, 23(3): 277–95.

Weinstein, J. M. (2007). *Inside Rebellion: The Politics of Insurgent Violence*. Cambridge University Press, Cambridge.

Wilde, M. and Williamson, J. (2016). Evidence and epistemic causality. In Wiedermann, W. and von Eye, A., editors, *Statistics and Causality: Methods for Applied Empirical Research*, pages 31–41. Wiley, Hoboken, NJ.

Williamson, J. (2005). *Bayesian Nets and Causality: Philosophical and Computational Foundations*. Oxford University Press, Oxford.

Williamson, J. (2006a). Causal pluralism versus epistemic causality. *Philosophica*, 77: 69–96.

Williamson, J. (2006b). Dispositional versus epistemic causality. *Minds and Machines*, 16: 259–76.

Williamson, J. (2007). Causality. In Gabbay, D. and Guenthner, F., editors, *Handbook of Philosophical Logic*, vol. 14, pages 95–126. Springer, Dordrecht.

Williamson, J. (2009). Probabilistic theories. In Beebee, H., Hitchcock, C., and Menzies, P., editors, *The Oxford Handbook of Causation*, pages 185–212. Oxford University Press, Oxford.

Williamson, J. (2011). Mechanistic theories of causality. *Philosophy Compass*, 6(6): 421–47.

Williamson, J. (2013). How can causal explanations explain? *Erkenntnis*, 78: 257–75.

Williamson, J. (2021). Calibration for epistemic causality. *Erkenntnis*, 86(4): 941–60.

Williamson, J. (2022). A Bayesian account of establishing. *British Journal for the Philosophy of Science*, 73(4): 903–25.

Woodward, J. (2003). *Making Things Happen: A Theory of Causal Explanation*. Oxford University Press, Oxford.

Index

For the benefit of digital users, indexed terms that span two pages (e.g., 52–53) may, on occasion, appear on only one of those pages.